basic plumbing and central heating

roy treloar

Launched in 1938, the **teach yourself** series grew rapidly in response to the world's wartime needs. Loved and trusted by over 50 million readers, the series has continued to respond to society's changing interests and passions and now, 70 years on, includes over 500 titles, from Arabic and Beekeeping to Yoga and Zulu. What would you like to learn?

be where you want to be with **teach yourself**

Orders: please contact Bookpoint Ltd, 130 Milton Park, Abingdon, Oxon OX14 4SB. Telephone: +44 (0) 1235 827720. Fax: +44 (0) 1235 400454. Lines are open 09.00–17.00, Monday to Saturday, with a 24-hour message answering service. You can also order through our website www.hoddereducation.co.uk

British Library Cataloguing in Publication Data: a catalogue record for this title is available from the British Library.

Library of Congress Catalog Card Number: on file.

First published in UK 2008 by Hodder Education, part of Hachette Livre UK, 338 Euston Road, London, NW1 3BH.

This edition published 2008.

Typeset by Transet Limited, Coventry, England.
Printed in Great Britain for Hodder Education, an Hachette Livre UK Company, 338 Euston Road, London NW1 3BH, by Cox & Wyman Ltd, Reading, Berkshire.

Impression number 10 9 8 7 6 5 4 3 2 1
Year 2012 2011 2010 2009 2008

contents

introduction

This book has been written with the domestic homeowner in mind. It identifies the plumbing systems in your home and explains how to undertake some basic plumbing work yourself. You might be considering the replacement of extensive pipework, or perhaps you'd just like to know what to do in an emergency: in any case this book will give you insight into the activities that a plumber might undertake should they be called upon, and provides you with clues about what they might need to do and why. It will also provide you with key questions to ask when seeking the services of a plumber.

The opening chapters give you information about the water supply pipe: how it enters the building from the street outside and travels through your house. Dealing first with the cold water supply, then the hot, you will find out how the heating works and also about the drainage of water from your property. As the book takes you around your home it identifies the main variations of plumbing systems found, and shows you specific things to look out for in the design, thus hopefully ensuring a trouble-free existence.

Chapters 03 and 04 will be key to those who wish to tackle simple repairs and find out what action to take in an emergency. You will discover how to deal with a collection of problems: from dripping taps and overflowing cisterns to blocked sinks and toilets.

Further chapters go on to discuss plumbing practices: identifying materials used, jointing methods and specialist plumbing tools and so on. This will help you to understand how to complete some of the work yourself. Larger plumbing and maintenance works that you might consider are also discussed, designed to ensure you are not faced with an emergency call-out from a plumber.

When faced with the subject of plumbing for the first time in your life there is always the fear of water pouring through the ceiling if you tackle the work yourself. This does not need to be the case. Plumbing activities generally follow basic principles that most people can undertake. Unfortunately with plumbing, there is quite a list of jargon that puts some people off before they start. If you're faced with an unfamiliar term check the glossary at the back of this book, which might shed some light on the problem you are trying to solve, making it all seem so much clearer.

The book is limited in terms of what it can cover in sufficient depth, therefore it must be understood that you should not attempt any work that might put you at risk, for example when working with the electrical supply to a particular component, such as the pump or immersion heater, which is mentioned within this book. There are several fundamental aspects of electrical safety, beyond the scope of this book, that must be undertaken when working on electrical supply systems because you could put yourself or others at risk of electrocution. The book discusses aspects of the gas installation; again if you don't fully know what you are doing it could prove fatal. In conclusion if you are not fully competent you should leave well alone, if in doubt call in an expert!

Some of the work that you decide to have done, either by yourself or by other persons, may be subject to legislation such as the Building or Water Regulations that are in force. When you call in a plumber you assume that they are competent and in compliance with these laws, unfortunately this is not always the case. You are generally none the wiser and possibly don't really care, just being happy to see the job done. I must point out that approval may be required if you're considering works involving new additions to your home. I recommend you read Appendix 1 relating to work affected by legislation.

Hopefully you can find the faith to tackle some of the smaller jobs yourself, you might surprise yourself and gain the confidence to tackle much bigger tasks in the fullness of time. With the escalating cost of calling in a plumber today, you should get your money back on the first successfully completed activity. I hope this book brings to you some happy plumbing results.

Good luck!

Roy Treloar

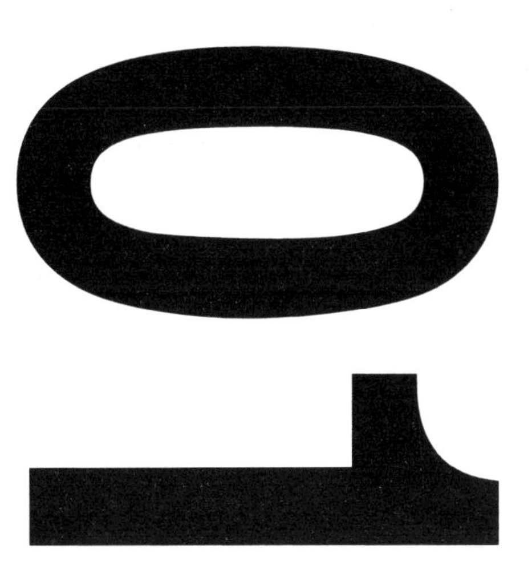

01 the plumbing in your home

In this chapter you will learn:

- how the water supply arrives in your home
- about the types of water supply
- about cold water storage
- about WC systems
- how water leaves the property.

This chapter looks at the plumbing systems in your home, from the point where water is fed into the house and passes through the pipework, to the point where the used and unwanted water leaves the house via the drains.

Incoming cold water supply

The water pipe feeding into your home comes from a supply pipe in the road, at a point just outside your property. There is usually a water authority valve at this point, and it is here that your responsibility for the water and pipework begins. The pipe travels below ground at a minimum depth of 750 mm to ensure that it is protected from damage and that the water will not freeze if the temperature drops below freezing point (0°C). The pipe then passes into a pipe duct through the foundations and ground floor into your home, terminating with a stopcock (tap).

In newer buildings a water meter will be incorporated within this supply pipe. This may be contained within a chamber outside, keeping the meter below ground level, or within the building itself, thereby allowing easier access for reading and maintenance. There may also be a stopcock situated underground at the boundary to your property, in addition to the one inside.

The pipe in the road from which this drinking water supply is taken is usually referred to as the 'mains'.

The water supply pipe

For the last 30 years or so, plastic (polyethylene) has been used for the cold water pipe feeding your home. Today it is typically blue and the standard diameter is 25 mm (which is equivalent to a copper pipe of 22 mm diameter) and is adequate to supply several outlets at once. In the past, however, smaller sized pipes were used, including:

- 22 mm plastic pipe – either black or blue (equivalent to 15 mm copper pipe size)
- 15 mm copper pipe
- 1/2 inch galvanized mild steel pipe
- 1/2 inch lead pipe.

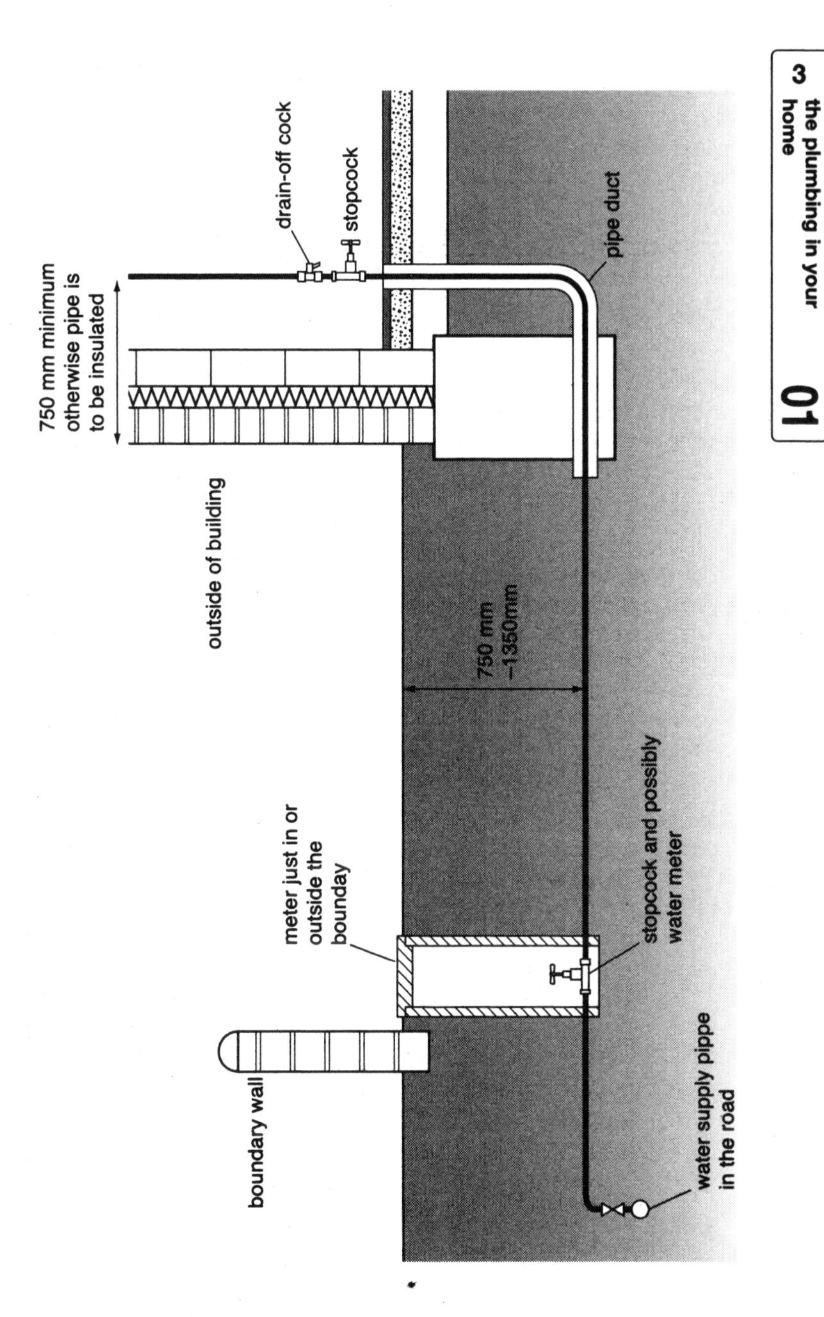

figure 1.1 cold water supply into a home

These older pipes are regarded as too small in a modern house because of the extra appliances used (washing machine, showers, etc.) and extra toilets. The size can restrict the flow of water and cause a loss of water flow at some outlets if several appliances are opened at the same time. Unfortunately there is not a lot you can do with your existing supply pipe if it's too small, other than replacing it with a new pipe.

Supply stopcock (stoptap)

It is very important that you know the location of this valve, after all, it supplies the water to the building and turning it off will stop the flow of water. This is essential in a situation where water is leaking from pipework. Typical locations for the stopcock inside the building are:

- under the kitchen sink
- in a downstairs toilet
- under the stairs, in a cupboard
- in a garage
- in the basement
- under a wooden floorboard, just inside the front door.

There may be an additional stopcock outside the building. Don't turn off this valve until you fully understand the consequences of doing so, as will be discussed in Chapter 02.

Ideally, once the internal stopcock has been found a label should be tied to the operating handle, so that anyone needing to find it in the future will know that this is the main water inlet to the building.

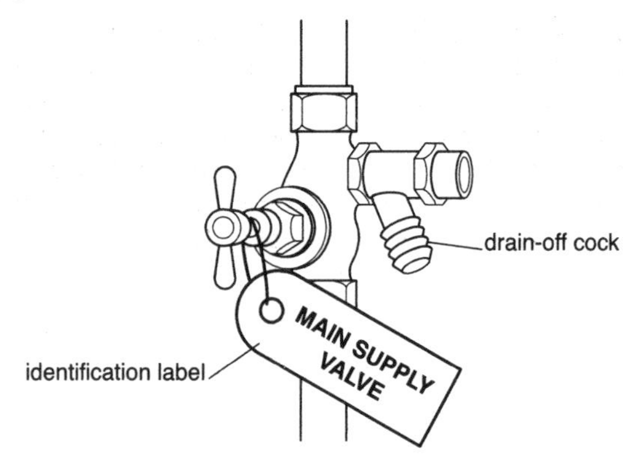

figure 1.2 supply stopcock with drain-off valve

Cold supply inside the dwelling

Once you have identified the incoming supply, look for a small outlet valve, known as a drain-off cock, just after the stopcock or incorporated within its design. This may be missing in older buildings or poorly installed systems. The drain-off cock allows the cold water supply mains pipework to be drained, for example for maintenance work or if you're going away for a long period of time in winter. There is the provision for a hose connection, but generally when the supply has been shut off much of the water can be drained out via the kitchen sink, so only that remaining in the pipe needs to be drained.

From the stopcock the pipe will run to the kitchen sink and other outlets. The route will depend on the system design, which will be one of the following:

- direct cold water supply
- indirect cold water supply
- modified cold water supply.

The pipework usually runs beneath floors or through pipe ducts, for example alongside the vertical soil or drainage stack (the drainpipe taking waste water from the building) as it passes up through the building. It may also be encased within the plaster wall. In all cases the actual pipe route is not a major concern provided that it is protected from unforeseen damage and frost.

Direct cold water supply

If you have this system, all of your cold water outlet points are fed directly from the mains supply. These include all appliances such as the sink, bath, basin and WC, plus any other outlets to washing machines, dishwashers or outside taps used for watering the garden. The cold supply may also feed a hot water system such as an unvented domestic hot water supply or combination boiler, discussed in Chapter 02.

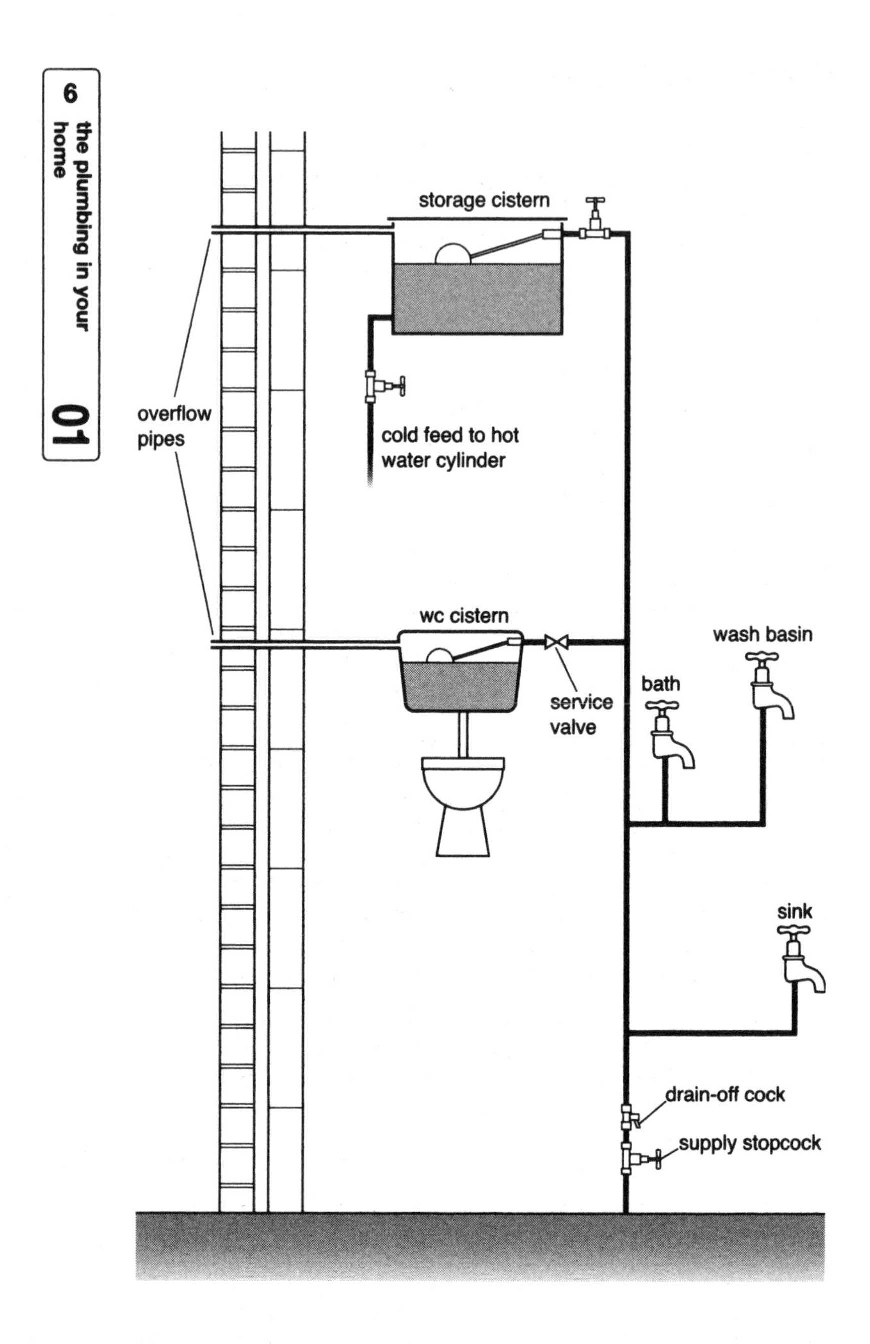

figure 1.3 direct cold water supply system

Indirect cold water supply

In this system the only appliance fed directly from the mains supply is the kitchen sink, plus a water softener if one is incorporated within the property. Instead of feeding directly to the other appliances, the supply feeds a water storage cistern, usually found within the roof space (loft). All other outlet points in the building are then fed from this storage cistern (see figure 1.4).

Modified cold water supply

This type of system is a combination of both the direct and indirect supply systems. In other words, there may be several outlets from the mains supply and several fed via a storage cistern.

Prior to the 1980s most systems were of the indirect design. These were designed to maintain a flow of water under the worst possible conditions, such as when the supply was cut off for some reason, such as being closed by the water authority for essential repairs, or in areas where there was an excessive drop in water pressure at peak times.

The local water authority may also have imposed a specific requirement that the supply had to be of an indirect design. However, today, due to higher pressures and consumer demand for combination boilers, unvented hot water supplies and guaranteed availability of drinking water, more and more systems rely on direct mains supply pipework. Also, with the direct system supplying both cold and hot supplies there is no need to have a cistern in the roof space or to extensively insulate the pipework and cistern from freezing up in the winter.

It is important to note that where all outlets are supplied via the mains supply, the supply pipe must be of a sufficient size (minimum 22 mm), otherwise some outlets will be starved of water when several outlets are open at the same time, as mentioned earlier.

What outlets are fed directly from the supply main?

To find out which outlets are fed directly from the cold mains supply pipe in your home is a simple process. First, turn off the incoming stopcock (see page 4) and then go around to all outlet points (taps) on the system to see which do not have any water

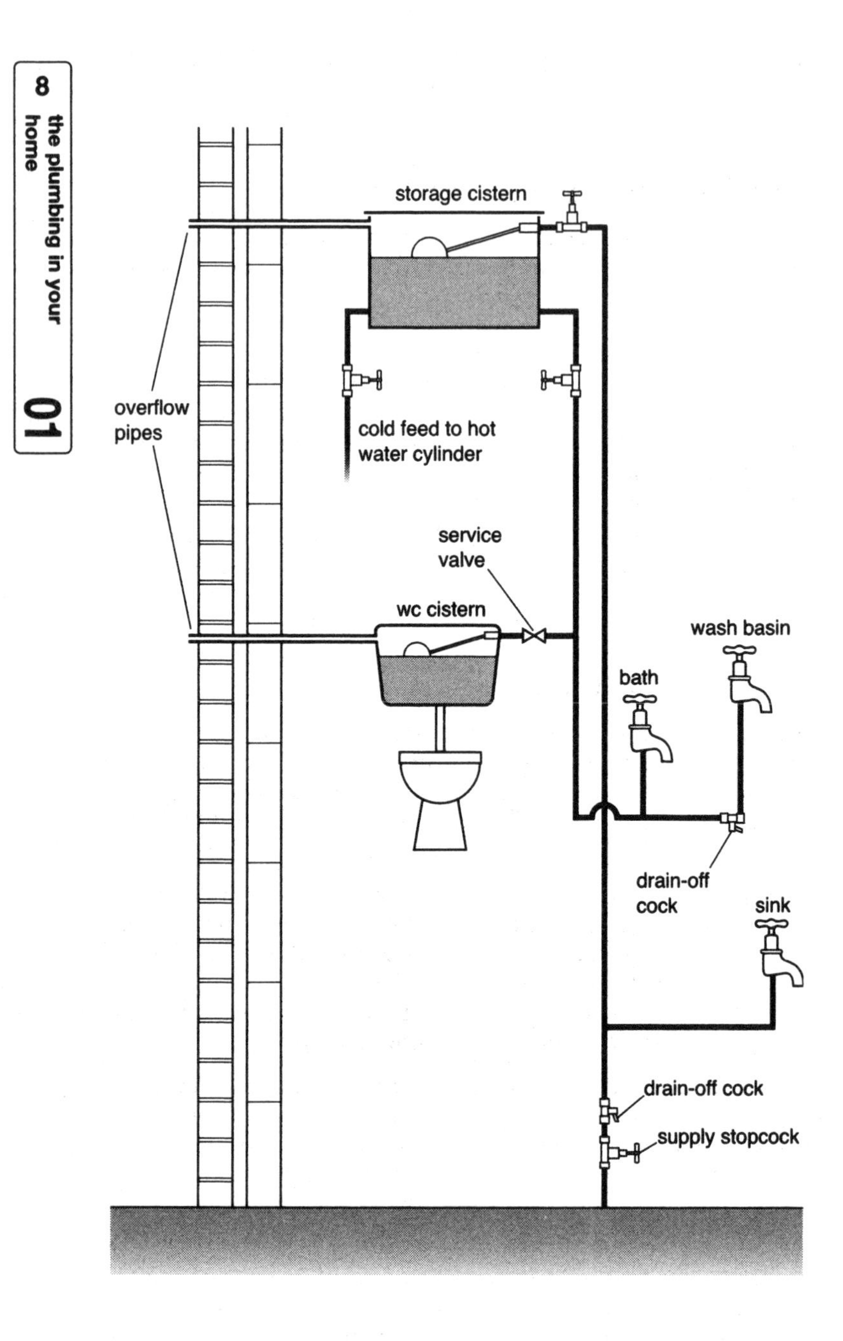

figure 1.4 indirect cold water supply system

flow when the tap is turned on. Likewise, to find out if the toilet cistern is fed from the mains supply, flush the toilet to see if it refills.

Drinking water (potable water)

It may be a surprise to learn that with modern systems, if designed and installed correctly, all outlet points, both hot and cold, should be supplied with water fit for human consumption, even where supplied via a cistern in the roof space. When we look at the installation of the pipework and appliances you will learn that the water is protected from contamination at all costs. For example, in figure 1.5 you will see that a filter has been incorporated within the overflow and that the cistern itself has a tight-fitting lid with all connections designed to prevent anything getting in and contaminating it, such as insects. So, water that has been stored in a cistern will also be regarded as safe to drink, so you must ensure under all circumstances that it remains this way.

The storage cistern

Figures 1.3 and 1.4 show a cistern that contains a large volume of water for the purpose of supplying hot or cold water pipework that is not fed directly from the mains. Buildings in which everything is fed directly from the cold mains water supply do not have a storage cistern.

The water level inside the cistern is regulated by the use of a float-operated valve, designed to close off the water supply when the desired water level is reached. Should this valve fail to operate, the water will continue to rise until the overflow pipe is reached, at which point it will overflow, warning the occupants of the building that something is wrong.

For the last 35 years or so storage cisterns have been made of plastic materials, however, some very old galvanized cisterns can still be found. Where this is the case it may be worth considering a replacement as it might have exceeded its expected lifespan.

All storage cisterns fitted since 1991 should be of a design that incorporates a tight-fitting lid and filtered overflow to ensure that not even the smallest of insects can get in to contaminate the water supply. Even the vent pipe from the hot water supply (discussed later) passes through a rubber grommet in the lid.

Around all this is a snugly fitted insulation jacket, and all the pipework to and from the cistern should be similarly insulated. Older installations may not be protected to such a high standard, and if an inadequate system is encountered (for example, with a loose or flimsy lid) the water should be treated with caution where it is used at cold or hot water outlets. If there is no lid at all, this needs to be remedied immediately. Dead bats are commonly found floating and rotting in unprotected cisterns.

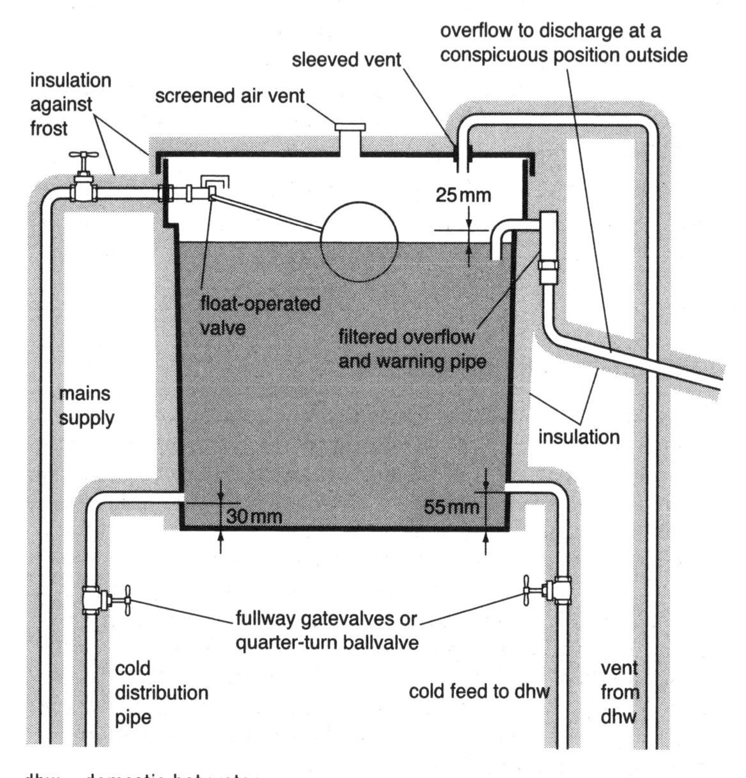

dhw = domestic hot water

figure 1.5 cold water feed and storage cistern

The condition of the storage cistern needs to be inspected occasionally to check that it is sound and protected. Ideally, once a year remember to check:

- the filters found in the overflow and lid to ensure they are not blocked, for example with flies
- the operation of the float-operated valve to ensure it is closing properly.

The float-operated valve (ballvalve)

The float-operated valve found within the storage cistern is generally the same as that found within a toilet cistern, although many of the newer toilet cistern float-operated valves are of a different design. The float-operated valve is often simply called a ballvalve, taking its name from the large ball float attached to the lever arm, which floats on the surface of the water. As the water inside the cistern rises and falls, so does the float.

These valves work on the principle of leverage, in that as the water rises the long arm lifts and forces a washer up against the water supply inlet. Generally only two designs of float-operated valve will be found, as shown in figure 1.6. The older valve, known as the Portsmouth ballvalve, can no longer be installed as it contravenes current Water Supply Regulations. There are two main reasons for this:

1. Its inlet will be submerged at times when the valve is overflowing. If you look closely at the two valve designs, you will notice that the Portsmouth valve lets water into the cistern from below the valve body, whereas the diaphragm valve lets water into the cistern from above the valve body. The advantage of discharging at the higher position is that it alleviates the problem of the valve outlet becoming submerged when the water level has risen in the cistern due to a faulty valve, and possibly overflowing. When the outlet is submerged in this way it is possible that under certain conditions, where a negative force is acting within the mains supply pipe, water could by sucked back into the supply, causing water contamination.
2. With the Portsmouth valve, in order to adjust the water level in the cistern you must bend the lever arm as necessary. The modern valve has an adjusting screw to make the appropriate adjustment to the water level in the cistern.

If you need to replace the float-operated valve for any reason, it is essential to replace it with the modern diaphragm type. Repair work on these valves is discussed in Chapter 02.

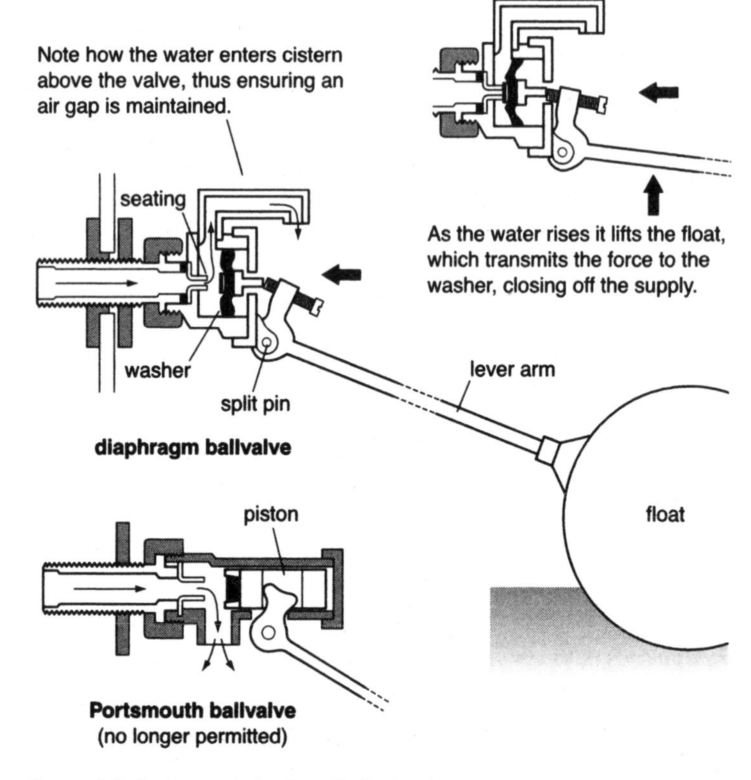
Note how the water enters cistern above the valve, thus ensuring an air gap is maintained.
seating
washer
split pin
diaphragm ballvalve
As the water rises it lifts the float, which transmits the force to the washer, closing off the supply.
lever arm
float
piston
Portsmouth ballvalve
(no longer permitted)

figure 1.6 float-operated valves (ballvalves)

Head pressure and flow

Finally, before we leave the storage cistern, we will consider the water pressure and volume of water flow that can be expected from the pipe supplying the water.

Pressure is the force of the water. Water pressure can be created by:

- a pump
- a storage cistern positioned high above the water outlets.

Flow is the volume or amount of water passing through a pipe. Water flow is dependant on the pipe size. A 22 mm diameter pipe will clearly allow a greater flow of water than one, say, 5 mm in diameter and consequently will fill up an appliance, such as a bath, much more quickly.

The cold water supply feeding your home will be supplied typically via a pump located at the water treatment works. This creates pressure within your supply pipe of up to around 3 bar (300 kN/m^2). However, when water has been stored in a cistern in your home, possibly located in the loft or roof space, its pressure is no longer influenced by the cold water mains supply but is now dependent on the position of the cistern. The pressure is considerably lower than that in the water mains supply pipe. For example, where a shower takes its water from a storage cistern, there might be only a two-metre head of water, in which case the water pressure will be so low that taking a shower is not practical. The term 'head' relates to the position of the water level in the system above the point where it is being drawn off, i.e. in the example below the water in the cistern is two metres above the shower.

There is a simple calculation that can be completed to find out the pressure created by an elevated cistern. This is: the head of water × 9.81. So, where the head is only 2 metres the pressure will be:

$$2 \times 9.81 = 19.62 \text{ kN/m}^2$$

This is about $^1/_5$ of a bar in pressure (100 kN/m^2 = 1 bar), and therefore far less than that expected from the mains supply pipe.

From this we can see that a storage cistern should be located as high as possible within a building. The pipe from the storage cistern needs to be a minimum diameter of 22 mm and, where several outlets are to be maintained, it may need to be increased to 28 mm or a second outlet may need to be taken from the cistern. Failure to observe these simple rules will mean that appliances are very slow to fill.

The toilet flushing cistern

The flushing cistern found above your toilet has undergone several changes over the last 15 years. The water supplied to the cistern is controlled by a float-operated valve. Most of these valves are of a similar design to those used in the cold water storage cistern (see figure 1.6). There are some different designs of valve, but these are beyond the scope of this book.

Prior to 1993, a nine-litre (two-gallon) flush was employed, and it had been like this since the toilet was first designed over 100 years ago. However, in order to try and conserve water this quantity was reduced first to seven and a half litres and then to a maximum of six litres, as per current regulations.

In order to discharge this water from the cistern into the toilet pan, a device is used that closes when the required volume has been discharged. Toilet cisterns traditionally worked using a siphonic device (see below), however, today there is another design which consists of a valve that is lifted to allow the contents to flow as necessary.

Flushing cistern operated by siphonic action

Siphonic action occurs where water is removed from a container, without mechanical aid, up and over a tube in the form of an upside-down letter J. The long leg joins to the flush pipe, the short leg is open to the water inside the cistern. If the air is removed from the tube a partial vacuum is created. This action, in the case of the flushing cistern, is triggered by the large diaphragm washer being lifted, which discharges a quantity of water over the top of the J-shaped siphon bend. As the water drops down through the flush pipe to the outlet it takes with it some of the air contained within, thus creating a partial vacuum. With the partial vacuum formed, gravity acts upon the surface of the water, pushing down and forcing the water up into the J-shaped siphon tube. As it reaches the top of the upturned bend it simply drops down to the outlet to be discharged into the pan, via the flush pipe. This action continues until the air can get back into the tube to break the vacuum and restore normal pressure. So, water will continue to discharge until the water level has dropped inside the cistern to that of the base of the siphon. The initial action of lifting the diaphragm washer is instigated by the operation of the lever arm located within the side of the cistern.

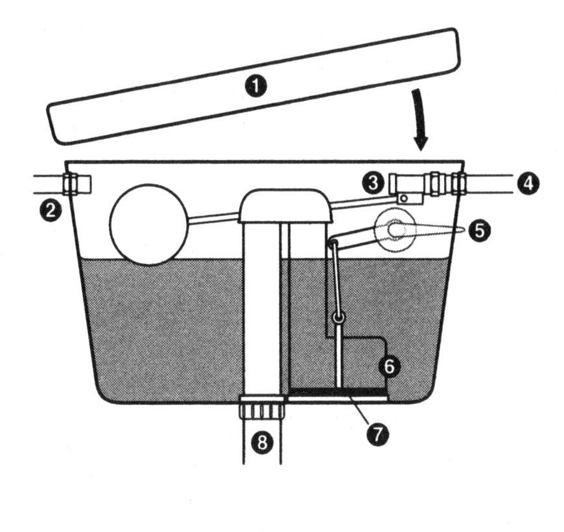

1 lid
2 overflow
3 float-operated valve
4 water supply inlet
5 operating arm
6 siphon
7 diaphragm washer
8 flush pipe

figure 1.7 flushing cistern operated by siphonic action

Valve-type flushing cistern

Several designs of valved flushing cistern have been developed within the last few years; the one shown in figure 1.8 works by allowing for a dual flush. A dual flush offers:

- a reduced flush for the purpose of removing urine from the toilet pan
- a full six-litre flush where there are solids to be removed.

There are two buttons housed within the cistern lid; one button has a shorter rod attached to it than the other. When the larger button, with the longer rod, is pressed it lifts the valve sufficiently to engage into a latch and is held up by a small float. Water now flows from the cistern and the latch only releases as the water level drops, taking the float with it. When the smaller button is pressed, the smaller rod does not lift the valve sufficiently to engage with the latch, so the valve is only raised for a short period while the button is held down. A linking cable operates a lever to initially lift the valve from its seating.

Note that a separate overflow pipe is not run from valved flushing cisterns because if the water level rises, due to the water inlet failing to close, it would overflow down through a central core, within the valve from the cistern into the toilet pan.

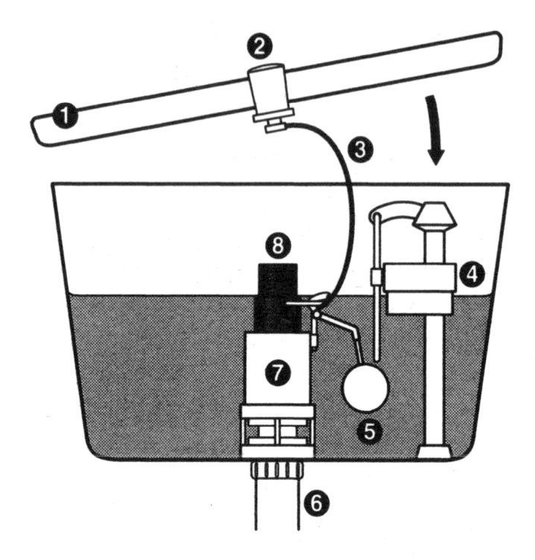

1 lid
2 push-button control
3 cable to operate valve
4 float-operated valve
5 float which holds valve up to provide full 6-litre flush
6 flush pipe
7 flushing valve
8 central overflow

figure 1.8 flushing cistern operated by flushing valve

Hard and soft water

Water is generally classified as being either hard or soft. This classification relates to the impurities that the water contains and is indicated as a measure of the number of hydrogen ions (acidic) or hydroxyl ions (alkaline) present in a sample of water. This is known as the potential of hydrogen value (pH value)

pH value of water

1 2 3 4 5 6 7 8 9 10 11 12 13 14

soft (acidic) ⟶ (alkaline) hard

neutral

Hard water contains calcium carbonates, sulphates or magnesium, which basically means limestone in one form or another, whereas soft water does not. This limestone has dissolved in the water because it is a natural solvent. The hardness of water can be further classified as:

- permanently hard (contains dissolved rock such as calcium or magnesium sulphate)
- temporarily hard (contains dissolved rock such as calcium carbonate).

The limestone in permanently hard water cannot be removed without water-softening treatment. Temporary hardness, however, occurs where the water has fallen onto calcium carbonate. This is a different form of limestone and will only dissolve in the water if it contains carbon dioxide, which the water acquired as it fell as rain. Boiling the water can remove the temporary hardness as the carbon dioxide can escape from the water; boiling will have no effect on permanently hard water.

Soft water, on the other hand, does not contain any dissolved limestone. It is more acidic or aggressive as a solvent, and will soon destroy metals, in particular lead when used within a plumbing system. Soft water feels different from hard water and is more pleasant to wash in; it is also much easier to obtain a lather when using soap in soft water, and takes longer to rinse the soap away. Hard water is also distinguishable by the scum that forms on the surface of the water and around sanitary appliances, and by the limescale that forms around taps and in the toilet bowl.

Limescale

Limescale can be found at the outlet point of both hot and cold taps in hard-water areas. What is it? In a nutshell, it is caused by temporarily hard water! However, a more in-depth explanation is appropriate here.

When it rains, the water falling from the sky is enriched with carbon dioxide (CO_2), trapping it within its molecular structure. This water falls to earth and percolates through the ground on its way to the rivers and reservoirs. If it flows through limestone during this journey, the CO_2 in the water causes the limestone to dissolve and, as a result, the limestone is carried in suspension within the water. When the CO_2 can escape from the water, such as by rapid shaking movements or heating the water above

60°C, the limestone will not remain dissolved, as it was the CO_2 that maintained this condition. Consequently, the limestone falls from the water and collects within the system as solid limestone (limescale). It is found around tap heads because the water collects here as it leaves the spout, and the CO_2 escapes back to the atmosphere as the water evaporates.

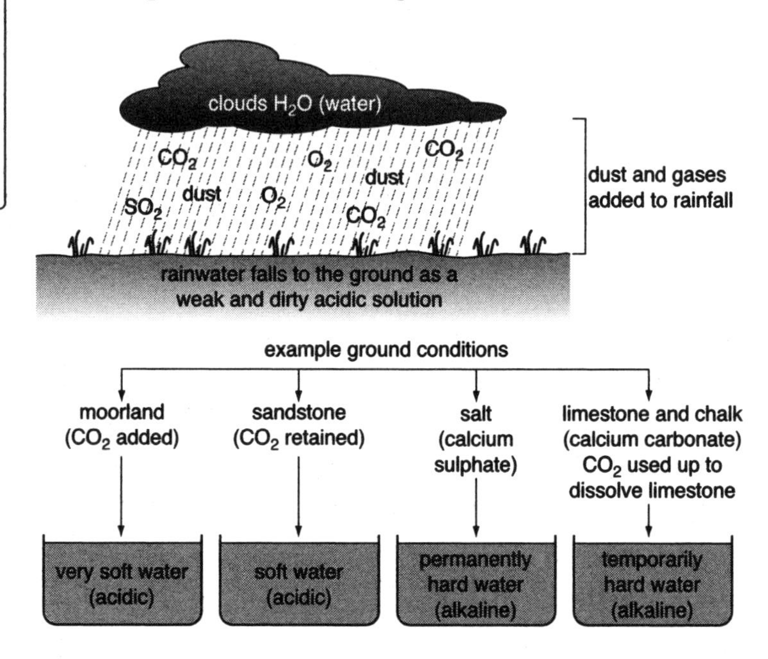

figure 1.9 formation of soft and hard water

It is important to note that limescale can collect, unseen, inside your pipework, accumulating around heating elements and heat exchanger coils, causing long-term damage and affecting heat-up times. It can drastically reduce the rate of water flow through any pipes in which it collects. To prevent this, it is essential to store the water at a temperature no higher than 60°C, as above this temperature the CO_2 can easily escape from the water. Limescale can also be prevented by using water softeners and conditioners.

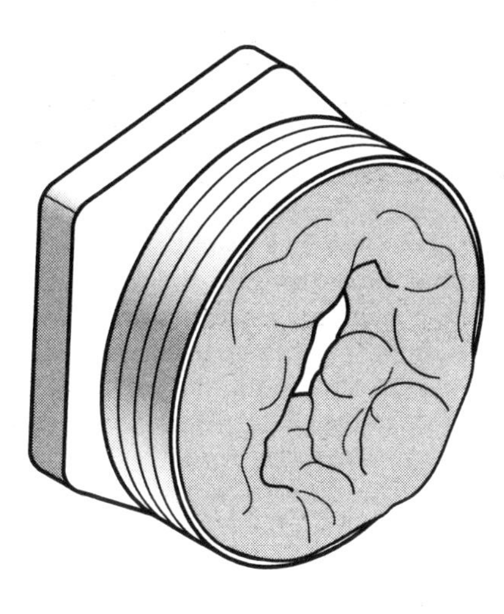

figure 1.10 limescale build-up in a pipe reducing the size

Water softeners and water conditioners

A water softener

This is a device that is designed to remove all of the calcium and magnesium ions (lime) from the water. Basically, when the water is passed through a special chemical bed called zeolite, or through very small plastic beads covered with sodium ions, the calcium and magnesium is given up. Eventually, the zeolite bed becomes exhausted to the point where it stops softening the water. It is then time to regenerate the bed material with sodium ions. This is achieved by passing a salt solution (brine) through the softener to displace all of the calcium and magnesium and recharge it with sodium. The regeneration process flushes out all of the unwanted products into a drain. The process of regeneration is completed automatically, timed to take place during the early hours of the morning; during this period no softening takes place and hard water will be supplied if a tap is turned on. A water softener is the only device that removes the calcium and magnesium from the water (see figure 8.2).

A water conditioner

This is not a water softener but a device that reconditions the small dissolved particles of limestone, referred to as calcium salts, held in suspension in the water so that they do not readily stick together to form noticeable limescale. If you viewed untreated hard water under a microscope, the calcium salts would appear star shaped, with jagged edges. It is in this form that they stick together. The water conditioner aims to take off these jagged edges so that they cannot easily bind together and instead they simply flow through the system. There are two basic types of water conditioner. First there are chemical water conditioners, which use crystals that dissolve in water and bind to the star-shaped salts, sticking in the crevices and jagged edges and having the effect of rounding off the sharp points. The other type of water conditioner passes a small electric current of a few milliamperes across the flow of water. This current alters the shape of the calcium salts, changing them to a smoother and more rounded shape. This current is often produced by a magnet, although other methods can be used.

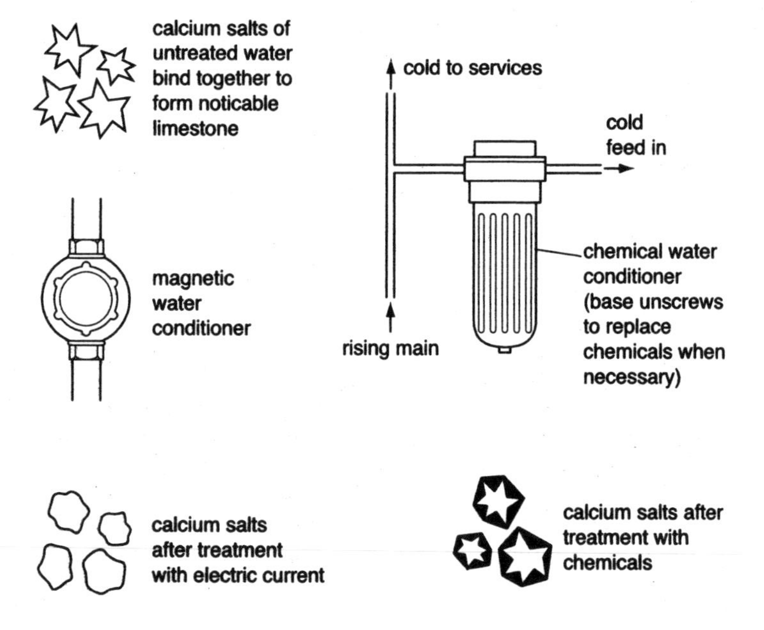

figure 1.11 water conditioners

The above-ground drainage system

The first thing water does as it goes down the plug hole is to pass around a range of bends that forms a small trap of water. You can see this trap by looking into your toilet pan or beneath the kitchen sink. Why is the trap there? It is not there to trap your wedding ring should it come off your finger, although this function does prove useful in such a circumstance. Its purpose is to provide a pocket of water between the outside air and the foul air within the drain and sewer. This air would most certainly be foul smelling but may also contain methane gas, which could prove hazardous. Another purpose of the trap is to prevent any vermin that may be in the drain from entering the building. This trap is the start of the house waste water system.

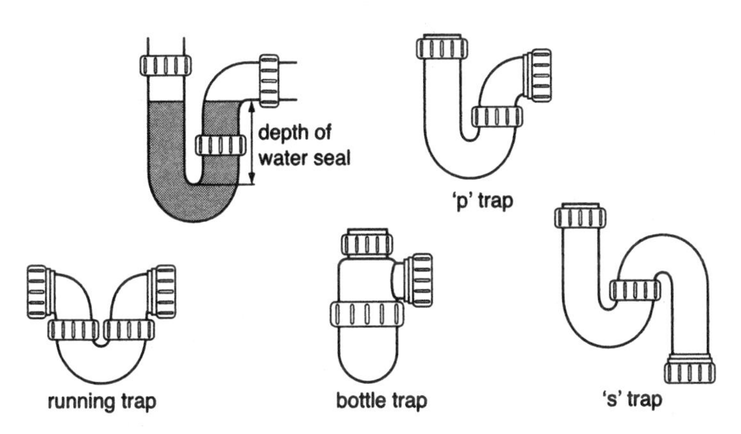

figure 1.12 traps

Gravity causes the water to flow from the trap along pipes that run down to adjoin the vertical discharge stack, referred to as the soil and vent pipe, and from here all the various waste pipes converge to take the fluid to the drainage system below ground. Obviously, the pipe must always be laid to fall in the direction of the water flow and the pipe must never, under any circumstances, be run uphill as water simply will not drain from the pipe.

The system illustrated in figure 1.13 is generally referred to as the single-stack system, although it is given the fancy title of 'primary ventilated stack system'. This has been installed in homes now for over 60 years.

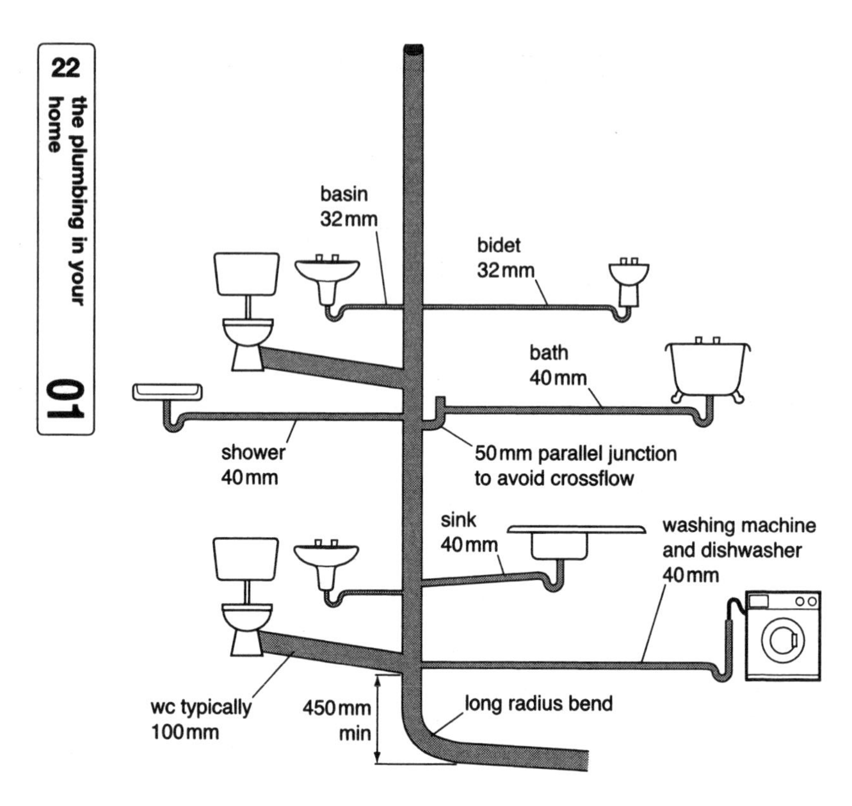

figure 1.13 typical primary ventilated stack (single-stack) system

There are many houses around that are much older than 60 years, therefore there are systems in existence, such as that shown in figure 1.14, which have a separate waste water discharge stack and foul-water stack. It was not until the pipes reached the ground-level drain that they were joined together. When major refurbishment to these antiquated systems is undertaken the plumber will update the system and install the single stack.

Plastic pipework is used for modern systems. This will either be of a type that can simply be pushed together, or the joints can be made using special solvent weld cement, which bonds the pipe to the fitting. The pipe diameters are shown in figure 1.13. The lengths of the pipes from the mains stack should be limited and this distance should not exceed the distances listed in table 1.1, otherwise you may experience problems with self-siphonage, explained below. It should also be noted that the flow of water

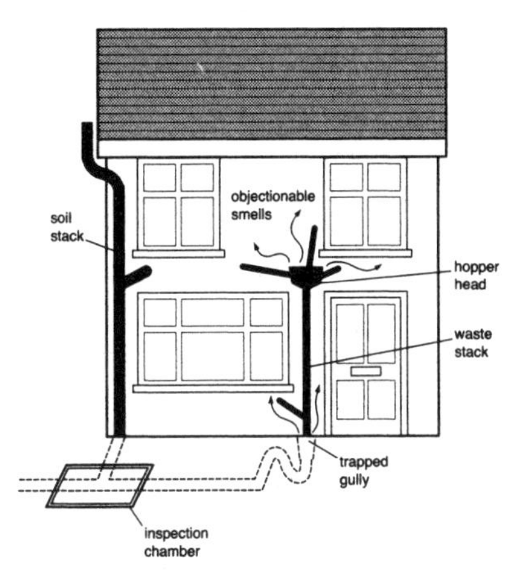

figure 1.14 the older system of separate waste stack and soil stack

passing horizontally to the vertical stack has been run to a minimal fall usually not exceeding a drop of between 18 mm and 90 mm per metre run of pipe. Exceeding this gradient could also create self-siphonage problems and can increase the problems of leaving any solid contents behind as the water rushes rapidly down the pipe.

table 1.1 maximum lengths for discharge pipes

Pipe size	Maximum length
32 mm	1.7 m
40 mm	3.0 m
50 mm	4.0 m
100 mm	6.0 m

Water siphonage from the trap

Water being siphoned from a trap is recognized by a gurgling sound coming from the appliance as air tries to enter the waste system in order to maintain the equilibrium of air pressure from inside the pipe to that of the surrounding atmosphere. Two types of siphonage can be encountered (see in figure 1.15):

- self-siphonage – caused as the water flows through the pipe forming a plug of water, causing a vacuum to be formed, sucking with it the water from the trap

- induced siphonage – occurs when no water has been discharged and is simply the result of a design error caused by the installer joining two waste pipes together so that as the water of one appliance flows past the branch connection of the other the air is drawn from the pipe.

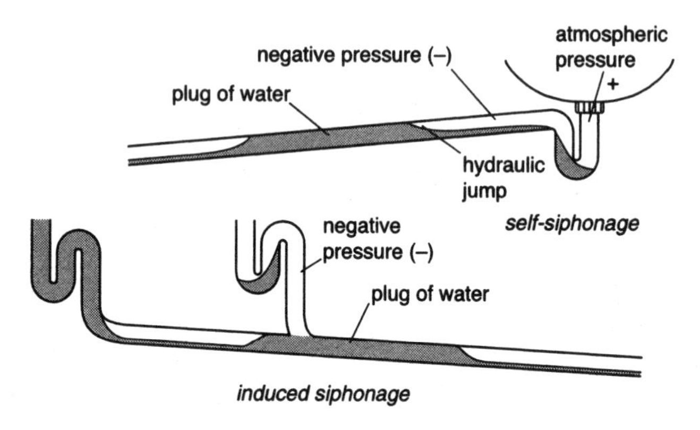

figure 1.15 water siphonage from the trap

Where continued problems are encountered with siphonage it is possible to fit:

- a resealing trap – this uses the concept of incorporating a non-return valve
- a special trapless (self-sealing) waste valve – sold under the manufacturer's trade name of a Hep_VO, this contains a special synthetic seal instead of the traditional water seal, which closes in the absence of water to seal off the pipe.

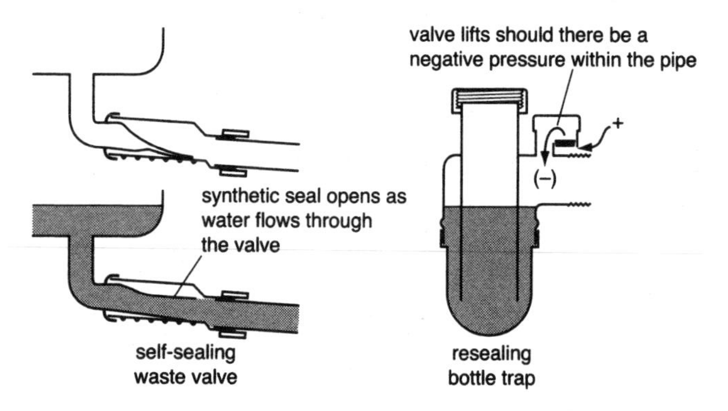

figure 1.16 alternative trap designs

Air admittance valves

Another device sometimes used to overcome problems with siphonage is an air admittance valve. This is basically like a big non-return valve that allows air into the drainage system but prevents air (potentially foul smelling) from coming out. So, where a negative pressure exists inside the drainage system these open in preference to the water being sucked from the trap.

Air admittance valves can be purchased in a whole range of sizes and sometimes the main discharge stack itself is terminated with an air admittance valve, possibly found within the roof space. This fitting is generally used where there are two soil stacks within the same building or where there are several buildings in close proximity. It overcomes the need to run the highest point of the discharge stack out through the roof, avoiding additional work to the roof tiles and ensuring rainwater cannot enter the building.

An air admittance valve must be fitted above the spill-over level of the appliance (the highest possible water level of the nearest adjacent appliance) otherwise, if there is a blockage in the pipe this fitting will be subject to a backup of water and the valve is unlikely to remain watertight.

Where these valves are in exposed locations, such as the loft, they do need to be insulated to ensure that they do not freeze up, as there is often a considerable amount of condensation within the pipe.

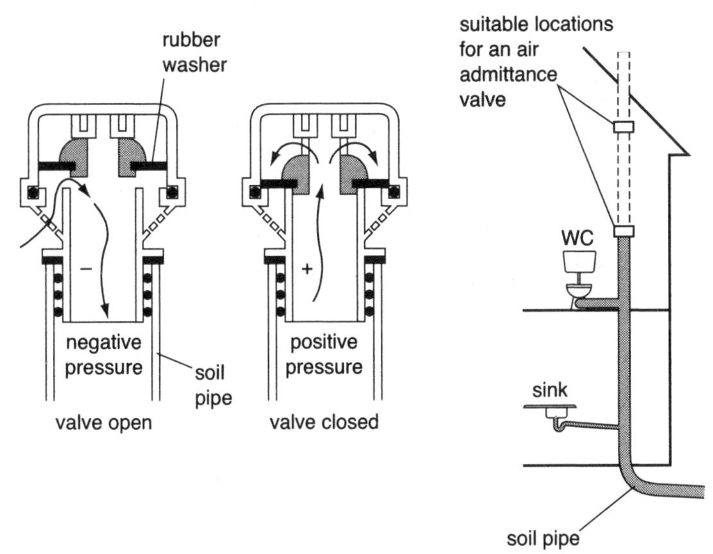

figure 1.17 air admittance valve

Access points

All good systems of drainage should have a means of access for internal inspection of the pipe, which is particularly useful when there is a blockage. Sometimes a large access point is positioned to the end of a small vertical section of 100 mm diameter discharge pipe, used as an alternative to the air admittance valve for an additional ground-floor toilet within the property. This method is acceptable provided that the pipe lengths are not excessive and, in all cases, no further than 6 m maximum from a ventilated drain, otherwise additional pressure fluctuation problems will be created within this section of pipe.

As with the air admittance valve, this access point must be installed above the spill-over level of the appliance otherwise, if there is a blockage to deal with, when it is opened the foul water will discharge all over the floor.

Remember that, if at any time you need to open an access point, you must consider what might lie behind – if water is there at a time of blockage, which may be the reason for opening this access point in the first place, it is likely to flow, with a surprisingly strong pressure, uncontrollably on to you and the floor where you are standing.

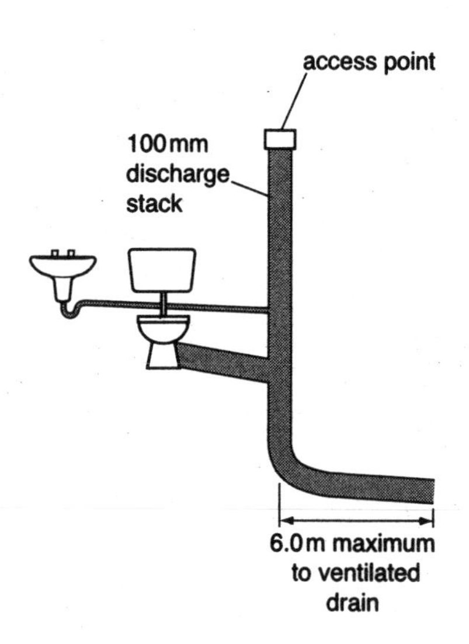

figure 1.18 access point

Pumped sanitation and drainage systems

For many years now there has been the opportunity to locate a drainage point for the purpose of removing water from basins, shower units and even from WC pan connections, from more or less anywhere within a typical house. These systems use what is called a macerator pump. This is basically a small holding tank, with the additional facility to macerate any solid matter within, which when full of water operates a pump to lift the water contents up or along through a small pipe (typically no bigger than 22 mm diameter) to discharge into a drainage stack.

The manufacturer's data sheets should be sought for the various design options, but typically the water could be elevated vertically by 4 m and horizontally the water could be discharged up to 50 m. One final point on the installation of these units is that it is a requirement that the property also has a conventional gravity system of drainage from a WC otherwise, if the power to the building is off due to a power cut, you would be without a toilet.

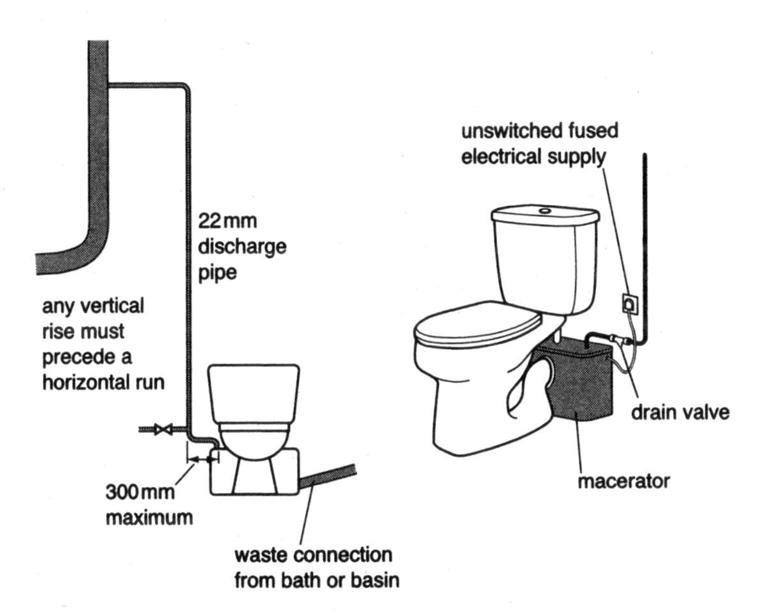

figure 1.19 pumped sanitation system

The water closet (WC)

The term water closet technically refers to the room into which a toilet pan is found. But when talking of the WC one is generally referring to the complete package of toilet cistern and attached pan.

The WC suite has undergone several design changes over the last few years. Today, Water Regulations limit the volume of water flushed down a newly installed toilet pan to a maximum of six litres, yet not many years ago this volume was nine litres. Most toilets installed these days are of the wash-down type, which basically means they rely on the discharging water flow to remove the contents from the pan.

Occasionally, siphonic WC pans will be found. These were installed quite extensively during the 1970s and are becoming quite rare these days as people update their homes. The siphonic pan, however, had one advantage over the wash-down pan in that it had the additional siphonic action to assist the removal of the pan's contents. It basically worked by lowering the air pressure from the pocket of air trapped between the two traps. This was achieved by allowing the flushing water to pass over a pressure-reducing fitting which created a negative pressure and sucked out the air between the two traps of water. With the partial vacuum created, the water and its contents in the upper bowl of the pan were sucked out by siphonic action. The cost of manufacture is possibly the reason for their disappearance.

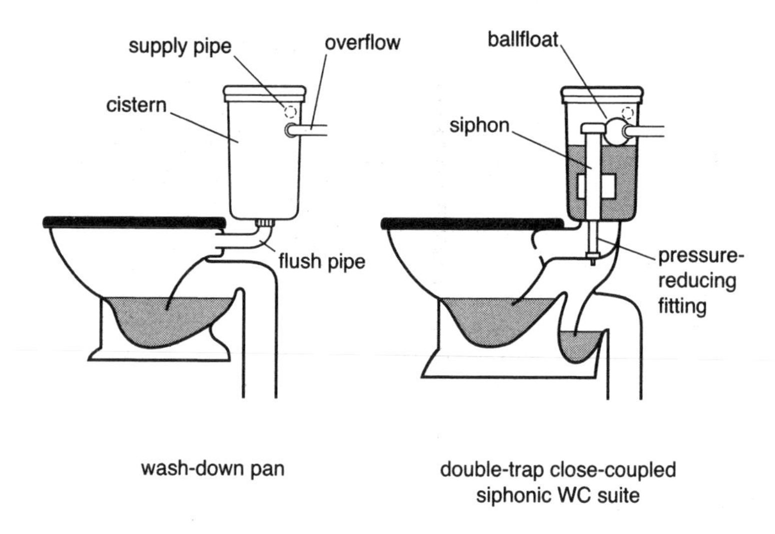

figure 1.20 WC installations

The below-ground drainage system

Once the water has reached ground level it is conveyed to the house drain, which removes it from the property to meet up with the public sewer, or the water may be collected in a septic tank or cesspit.

Septic tank

This is a private sewage disposal system used in some rural areas. Basically, all the foul and waste water is collected within a large double-compartment chamber, traditionally made of brickwork or concrete, although nowadays these are generally made from plastic. From here the water overflows through an irrigation trench to slowly filter into the ground away from the property.

These systems rely on a scum forming on top of the liquid and in so doing allow anaerobic bacteria to decompose most of the solids. Because not all the solids are broken down it is necessary to have the vessel emptied annually to remove the accumulation of the excess sludge that will not decompose. Failure to do so may lead to a blockage in the system.

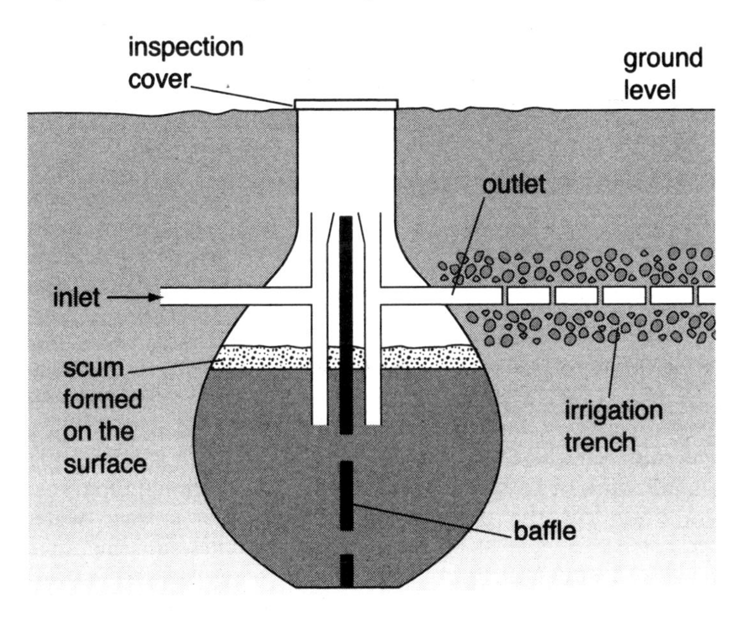

figure 1.21 septic tank

Cesspool (cesspit)

This is simply a watertight container that is used to collect and store waste and foul water from the property. Cesspools are used where no mains drainage has been connected to the property and there is insufficient provision for a septic tank. The tank will need to be emptied, ideally before it is full, by a contractor for proper disposal.

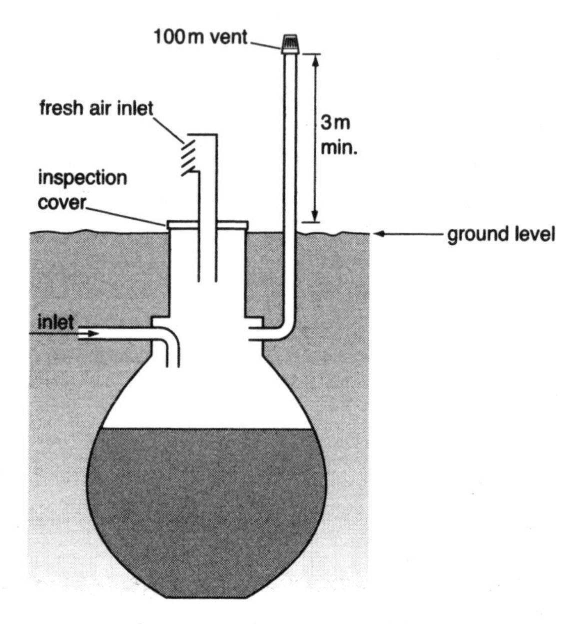

figure 1.22 cesspool

Surface water

In addition to the water that flows into the drains from the various sanitary appliances in the home, water is also collected from the gutters, rainwater pipes and large paved areas – this is generally referred to as surface water. If the drain is serviced by a septic tank or cesspool it will require an additional separately run drain for the purpose of collecting the surface water, because if this water is allowed to flow into these holding tanks it will cause them to fill too rapidly. In these cases, the surface water might be collected and run into a drainage ditch, river or soakaway.

The soakaway is simply a large hole filled with rubble, into which the drainpipe runs. The water collects here and gradually drains into the surrounding ground.

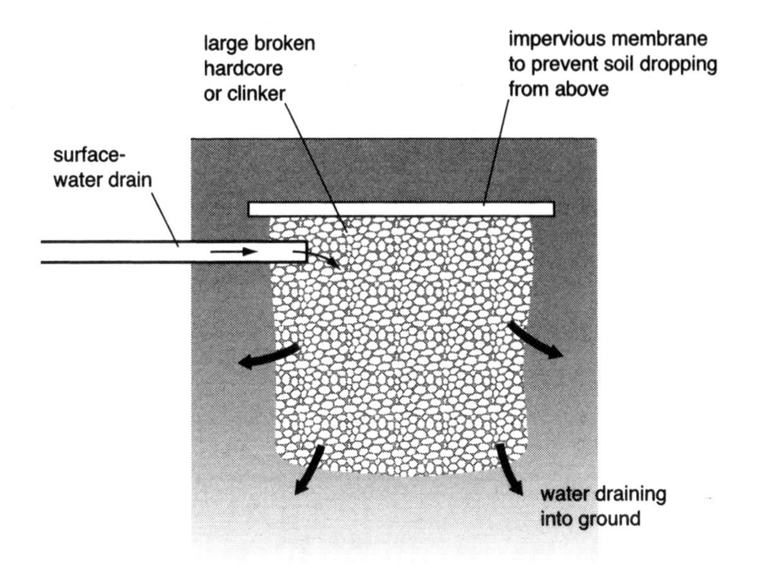

figure 1.23 the soakaway

Connections to public drainage systems

If the foul-water drain is connected to a public sewer, the surface water may be collected within the same pipe and run off from the property together. This is referred to as a combined system of drainage. Whether or not a combined system of drainage is used will depend very much upon the local authority, as it is they who treat all of the water. As a consequence, some areas have separate systems of drainage, in which the surface water is run into its own specific pipe.

When making any new connection to a drainage system it is essential to confirm the type of drainage system you have. Failure to do this could result in contamination of the local water course if you inadvertently discharge foul water into a surface-water drain.

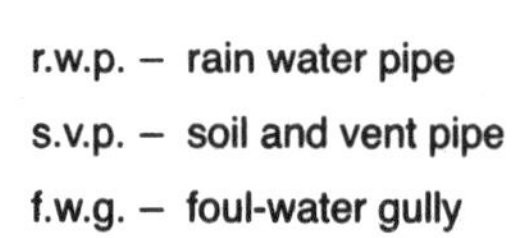

r.w.p. – rain water pipe
s.v.p. – soil and vent pipe
f.w.g. – foul-water gully

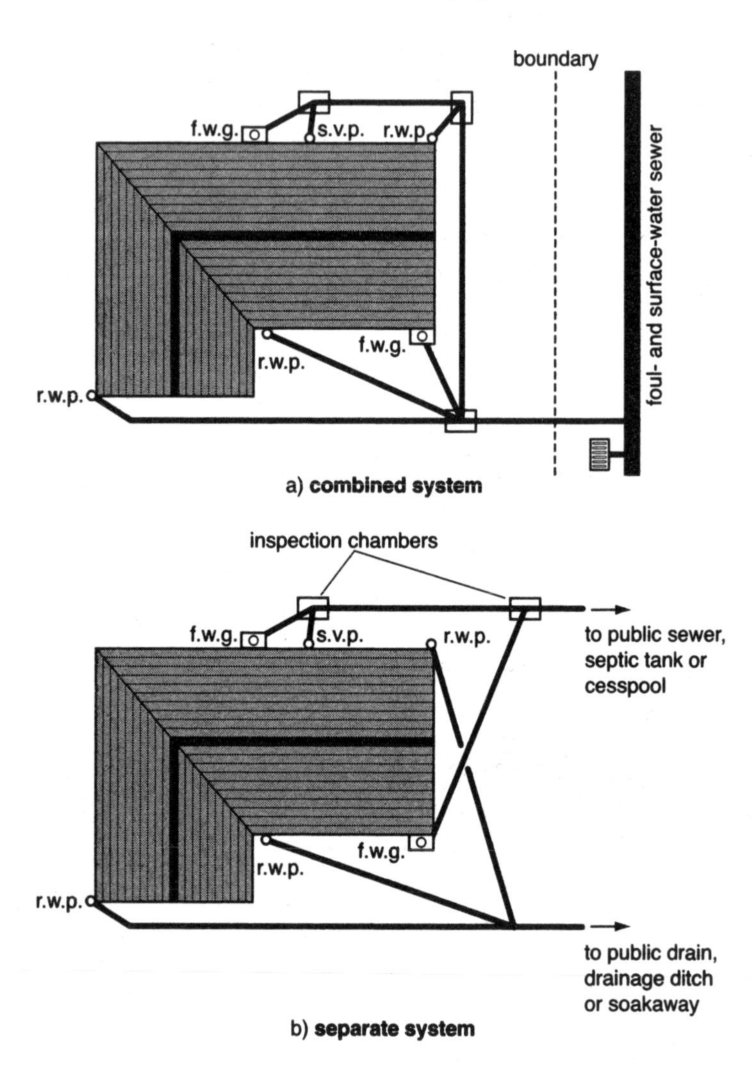

figure 1.24 connections to public drainage systems

In addition to the systems identified in figure 1.24, there is a slight variation that can be found where an odd surface-water connection is a long way from the surface-water drain, or there is some difficulty in passing the drainage pipe of a foul-water drain. In such a situation it is possible that this one-off connection can be discharged into the foul-water drain. If this is done, the system is referred to as partially separate, however, it must be understood that no cross-connection can be made the other way around, i.e. the foul water must never be allowed to connect to the surface-water drain.

Where a separate system of drainage is employed, the connections of the pipes to the surface-water drain do not have to include a trap. However, all connections made to the foul-water drain, be it surface water or waste water, must be trapped. If you look carefully at figure 1.24 you will see foul-water gullies (f.w.g.). These are traps at ground level, 100 mm in diameter,

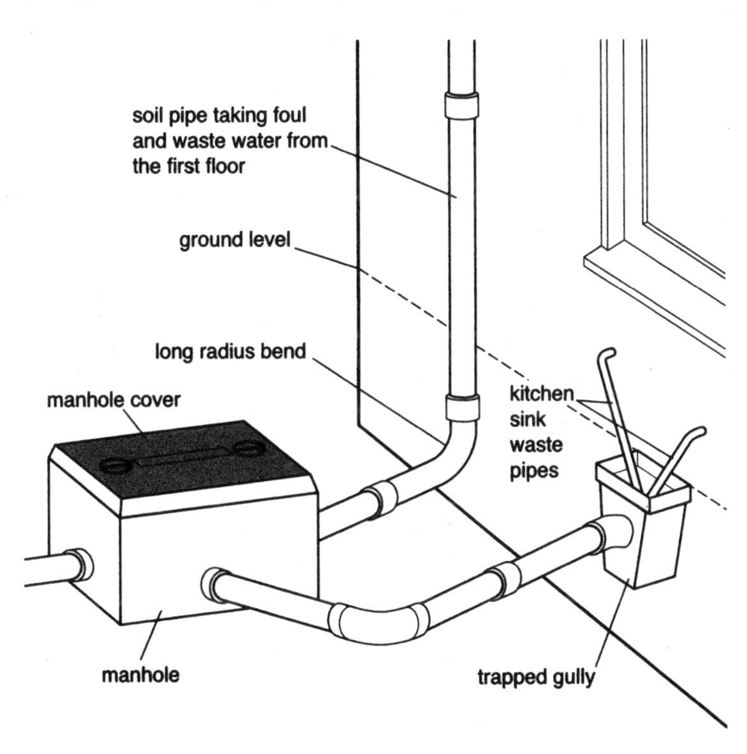

figure 1.25 connection of the above-ground drainage into the below-ground drainage system

i.e. the same size as the house drain, into which smaller pipes have been run. Prior to the 1970s these traps were left open, with a small brick course around the opening and a grate above the water level, however, nowadays the pipes entering these gullies are discharged below the ground surface into a side inlet pipe and an access cover is secured at ground level.

The soil and vent pipe connected to the drain is not trapped. However, it should be noted that all appliances connected to this pipe are themselves trapped. This pipe allows the free passage of air into and out of the drain, thereby maintaining equal air pressures within the drain and outside it. Air flowing through the drain also assists in drying out any solid matter left behind during flushing and as it dries it shrinks and is easily flushed away during the next discharge of water.

Gutters and rainwater pipes

This is the last part of the plumbing system, the 'outside plumbing', and the only part that if it was to leak would make little difference. The guttering consists of a simple channel located at the base of a roof to catch the run off of water. From here the water runs to the outlet and falls down the rainwater pipe to the surface drainage system below. Forty years ago metal was used for the installation of this last part of the plumbing system but, like so many things today, plastic has long since replaced these older traditional materials.

Protective equipotential earth bond

Look at your gas meter or incoming water supply and you may see a green and yellow wire connected to the pipe. This is called a protective equipotential earth bond connection.

Where a water, gas or oil pipeline comes into or passes out from the building there is the potential for stray electrical currents, resulting from faulty electrical equipment, to pass through as they flow down to earth. This can be dangerous for anyone touching the metal pipework itself. To ensure that these stray currents can flow safely to earth via a specifically designed electrical route, an equipotential bonding wire of 10 mm^2 minimum size is attached to the pipe at the point of entry/exit of the building and this in turn is connected to the main earth terminal at the consumer unit.

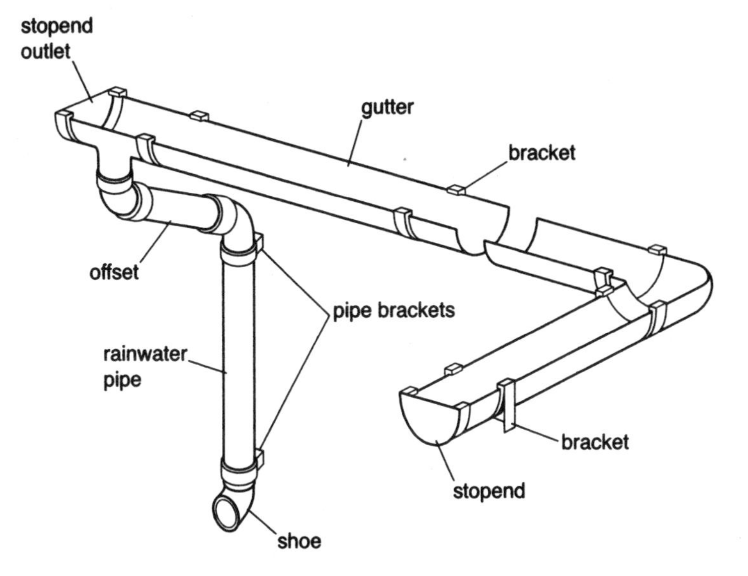

figure 1.26 gutters and rainwater pipes

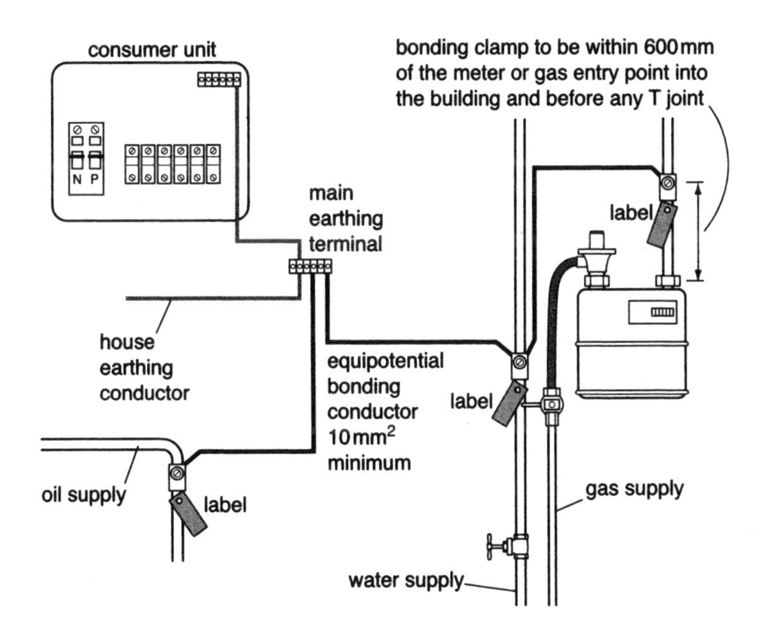

figure 1.27 equipotential bonding to all services entering the building

In addition to the bonding wires making a connection to the incoming services, there should be additional supplementary equipotential bonding wires linking together all the metalwork within wet areas, such as the bathroom. This will ensure that everything within that zone is at the same electrical potential, which is designed to prevent users receiving an electric shock.

On a point of safety, you should never disconnect these bonding wires without making sure that it is safe to do so. This may require the services of a qualified electrician for advice. See more about bonding in Chapter 06.

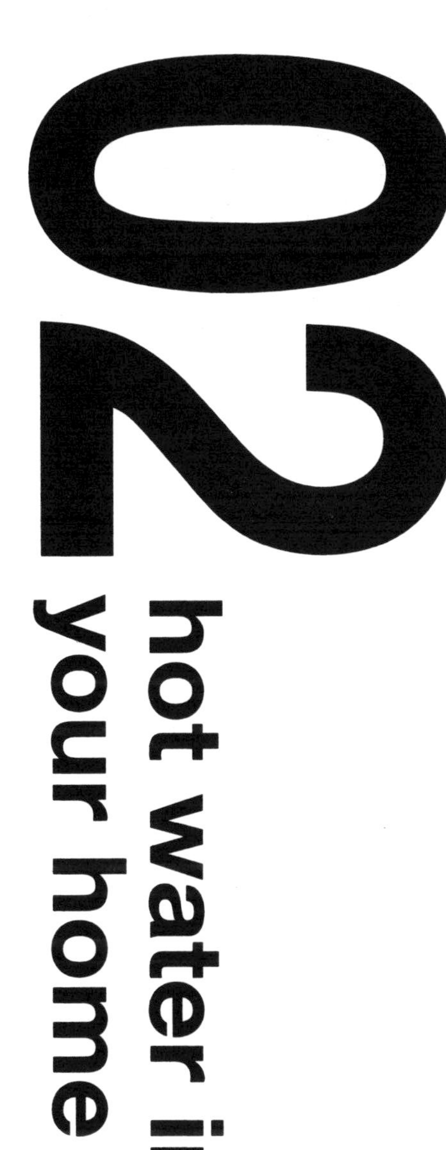

02 hot water in your home

In this chapter you will learn:

- **about the various fuels used to heat water**
- **about hot water supply systems**
- **about hot water storage systems**
- **how boilers differ.**

Gas installations

Many homes in the UK have a gas supply for the purpose of heating and cooking. The gas supply may be fed directly from the street outside your home to enter via a gas meter. Alternatively, you may buy your gas in bulk in a liquefied form and store it outside in a special holding tank until it is required, when it is drawn off automatically as it is converted to its gaseous form. The two methods of gas supply identified here are essentially irrelevant to you, the customer – you open a pipe and gas comes out.

The gas used in each case is slightly different and has different characteristics. For the most part you do not need to concern your self with the differences as they both burn in the presence of oxygen, producing a blue flame. These two gas types are called:

- natural gas – that fed directly from a pipe in the street
- liquefied petroleum gas (LPG) – that supplied in gas cylinders or bulk purchased.

Both of the gasses have a distinctive smell. This is not a true property of the gas, but a stenching agent which is added at the production plant so that it is easy to recognize should there be a leak. One of the key differences between the two different gasses is in the way that they react upon leaving the pipe. Natural gas is lighter than air so will rise upwards and is readily dispersed into the environment. LPG, on the other hand, is heavier than air and sinks downwards towards the floor and is not as easily dispersed, often gathering in low-lying pockets such as basements. LPG gas, when leaking from a pipe, drops around your feet and is not so easily smelt, which results in it rapidly accumulating undetected.

The gas pipework for a natural gas installation is fed through a gas meter, purely for billing purposes. Obviously this would not be required where you bulk purchase the gas.

It is important to note the location of the emergency control valve at the point of entry to the building. This should be accessible at all times so that if required the supply can be shut off very quickly. From this point the gas pipe is run to all the appliances requiring a gas supply.

Within a gas appliance, the gas is regulated and passed through a fine injector in order to allow the correct proportion of gas and air to mix within the combustion chamber where the fuel is burnt.

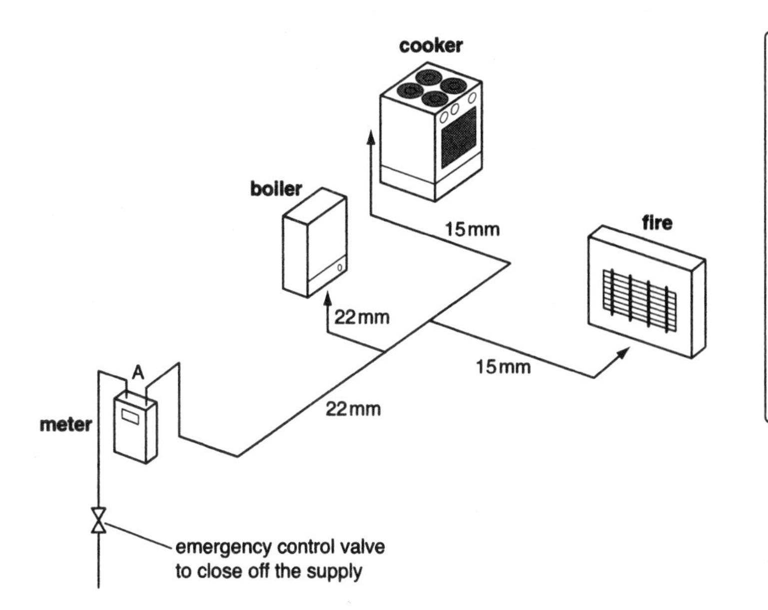

figure 2.1 layout of a typical gas installation

The appliance has many safety measures to ensure that gas will not flow through the appliance until it is required and that it can be burnt safely. Due to the potential danger of incorrectly installed gas fittings, the installation of pipework and the provision for its use fall under very strict regulations. It is not illegal to work on your own gas installation pipework or appliances on a DIY basis, but unless you are absolutely certain of what you are doing you would be ill advised to touch anything. Gas installers are trained and assessed to ensure their competency to carry out gas installation work; when any gas work is carried out within your home you must ask to see the engineer's CORGI card, identifying what areas of gas work they are allowed to undertake (see **appendix 1: legislation**).

A leak from a water-filled installation can cause a great deal of damage but rarely poses any real danger. On the other hand, a gas leak within a property is highly dangerous. When gas is burnt it is converted to water vapour and carbon dioxide, both of which are harmless gasses, being present in the atmosphere and within the air we breathe. However, if for some reason insufficient oxygen is available in the air used for the combustion process, incomplete combustion will occur and as a result carbon monoxide is produced.

Carbon Monoxide (CO)

Every year carbon monoxide gas poisoning claims the lives of around 30 people in the UK. Fuels that we burn (including coal, wood, oil and gas) are hydrocarbons, which are made up of hydrogen and carbon in various proportions. Both of these products can be burnt in the presence of oxygen and, if completely consumed, are converted to harmless carbon dioxide (CO_2) and water vapour (H_2O). However, if insufficient oxygen is available to support the combustion process, carbon monoxide (CO) will be produced because the carbon is not fully converted to CO_2.

Carbon monoxide does not have an odour and therefore cannot easily be detected. An appliance can discharge small quantities of these combustion products into the home without detection. Look at the symptoms in the chart below. These are common symptoms that are often simply attributed to stress or tiredness from overwork. If in any doubt, have your fuel burning appliances checked.

table 2.1 typical effects of carbon monoxide (CO) poisoning

Percentage of CO in the air	Symptoms/effects in adults
Less than 0.01	Slight headache after 1–2 hours
0.01–0.02	Mild headache plus feeling sick and dizzy after 2–3 hours
0.02–0.05	Strong headache, palpitations and sickness within 1–2 hours
0.05–0.15	Severe headache and sickness within half an hour
0.15–0.3	Severe headache and sickness within 10 minutes; convulsions and possible death after 15 minutes
0.3–0.6	Severe symptoms within 1–2 minutes and death within 15 minutes
1% or more	Immediate symptoms and death within 1–3 minutes

Note: Very small proportions of carbon monoxide within a room can prove fatal very quickly.

Oil installations

Some rural locations use oil as the source of fuel. Where this is the case the oil is supplied to the premises as a bulk purchase and stored in a large oil tank. Tanks today are generally made of plastic, and where a plastic tank is ordered to replace the traditional steel tank it is essential that there is adequate provision to support the entire surface area of its base, otherwise it will buckle and eventually split. Nowadays, where the oil tank is within a close proximity to the building, it needs to be of the bunded type. This means that there is a tank within a tank so that should a leak develop, the outer tank will contain the oil spillage.

An oil pipeline is run from the oil tank directly to the appliance. Oil burning appliances are generally limited to a boiler and sometimes a large range cooker. Along this pipeline several controls will be found, including:

- an isolation valve
- a filter
- a fire valve.

The fire valve is designed to close off the oil line in the event of a fire. Today these valves are installed outside, at the point of entry to the building, but in the past simple valves were installed within the appliance case.

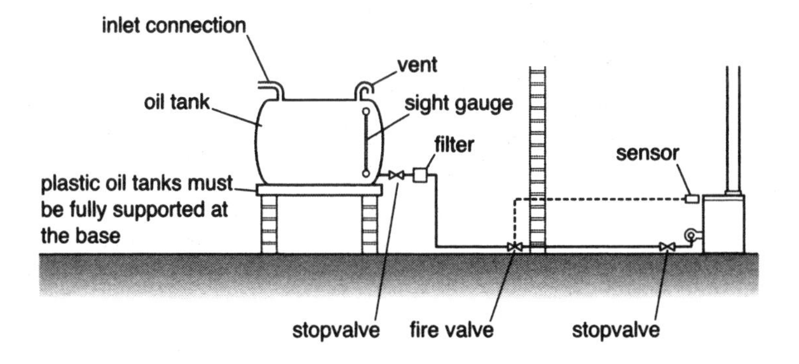

figure 2.2 oil line to boiler

Plunge a flame into a tank of oil and it will be extinguished. Oil needs to be atomized into a fine spray or vapour in order to burn. In modern boilers a pressure jet burner is used to consume the fuel. This forces the oil out through a fine nozzle where it is atomized and ignited within the combustion chamber of the boiler.

The cooker may also use this method or, alternatively, it may employ what is termed as a pot burner. This allows the fuel to flow slowly, driven by gravity, into a tray at the base of the burner. Here, the vapour is ignited and the flame passes up through the pot where it is mixed with the air supply to produce a safe, stable flame.

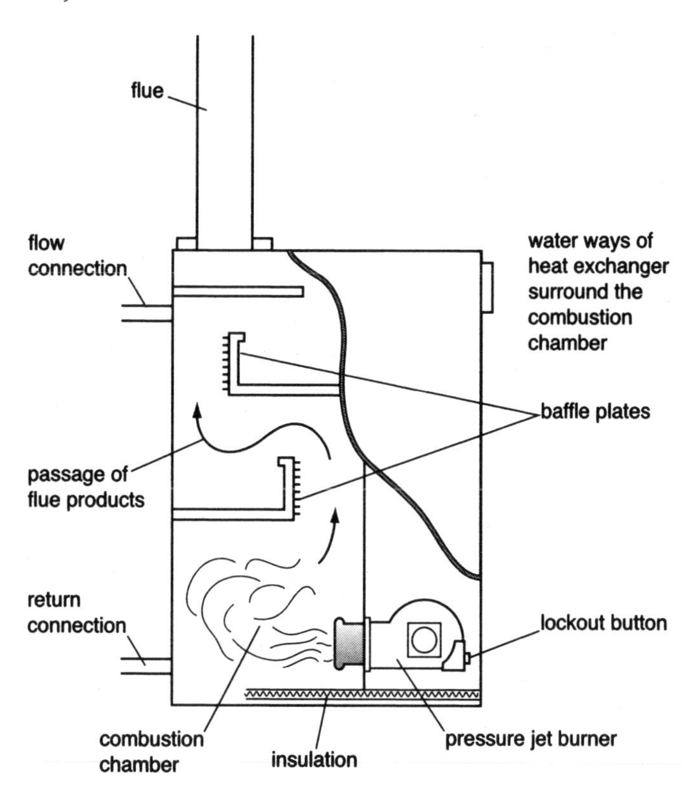

figure 2.3 oil boiler installation

Flues and ventilation for gas- and oil-burning appliances

Oil-burning appliances and many gas-burning appliances require the by-products of their combustion to be expelled out to the external environment. This is achieved by way of a flue pipe. The flue pipe of an appliance could be of several different designs – see Chapter 03. Air needs to be supplied in order to remove these by-products from the premises, otherwise the system will not work satisfactorily.

The installation of both gas and oil supply and the fluing and ventilation necessary for these systems is a very specialist subject and further reading is recommended for those with a particular interest in this area (see **appendix 3: taking it further**). Unfortunately, all too often the flue system or air supply requirement for appliances burning these fuels is not seen as important as the actual gas or oil supply pipework itself. Both these appliances can produce carbon monoxide (see above), which can be a silent killer in our homes.

Hot water supply

The design of your hot water supply will depend upon the location and age of your building. There are many variations of system design. The most common of these are:

- a gas or electric single-point water heater found above the sink or basin
- a gas multipoint water heater serving all hot water outlets
- a boiler used to store hot water within a cylinder; this system may also serve the central heating
- a combination boiler to heat both central heating and hot water instantaneously
- a thermal storage system (by far the least common).

These systems are classified as being either:

- storage (vented or unvented)
- instantaneous (combination boiler, multipoint, single-point or thermal storage).

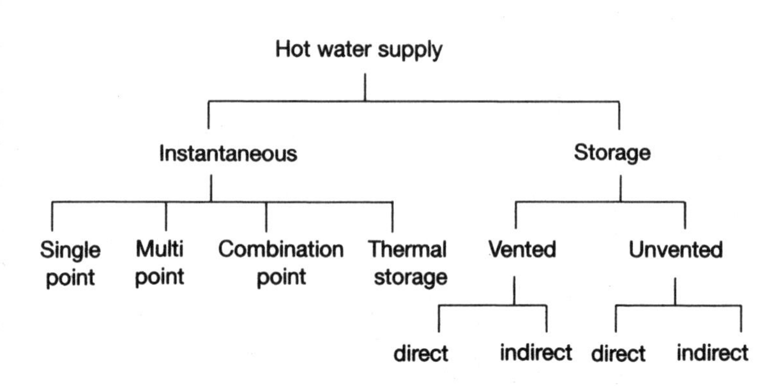

figure 2.4 types of domestic hot water systems

Hot water storage systems

Domestic hot water is stored in an enclosed vessel, which is most likely to be a cylinder, suitably insulated to keep the heated water warm. This vessel is found typically in an airing cupboard. The water is heated either directly or indirectly.

The installation of modern domestic hot water systems is controlled by legislation, which is particularly rigorous with regard to energy efficiency. If you want a new gas or oil boiler to use with a hot water cylinder, you cannot just install any old appliance. It must conform to the standards laid down within the Building Regulations, which are administered by the local authority. Consequently, when a boiler or cylinder is replaced, the local authority may wish to be notified in order to ensure that it is in compliance with current standards.

Storage cylinders have developed and become more efficient over the years. Older cylinders:

- required a cylinder jacket to be tied around them in order to keep as much heat as possible from being lost to the surrounding space. They were usually installed in a cupboard, which stayed warm and dry and thus provided an ideal storage area for airing clothes. However, in this modern

age of energy efficiency they have been identified as using fuel inefficiently
- had 1½–2 turns in the internal pipe coil that made up the heat exchanger. This led to a very slow heat transference rate and increased the time taken to heat the water in the cylinder as it passed from the primary heating circuit.

Modern cylinders:

- are foam lagged at the manufacturing stage
- have at least five to six turns in the heat exchanger, increasing heat transference times.

It is also possible to purchase high-performance cylinders that have a bank of many coils passing through the cylinder, allowing for even faster heat-up times.

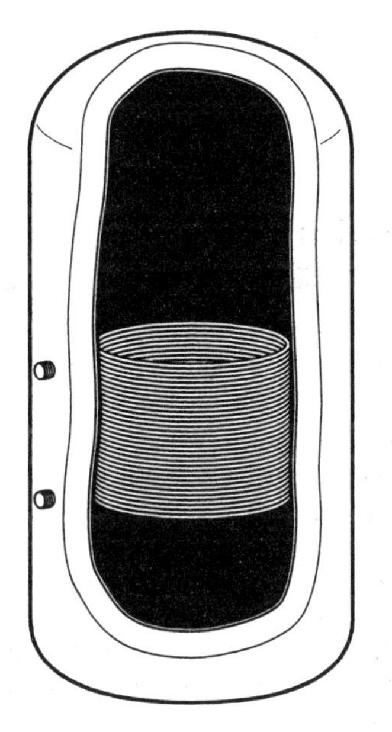

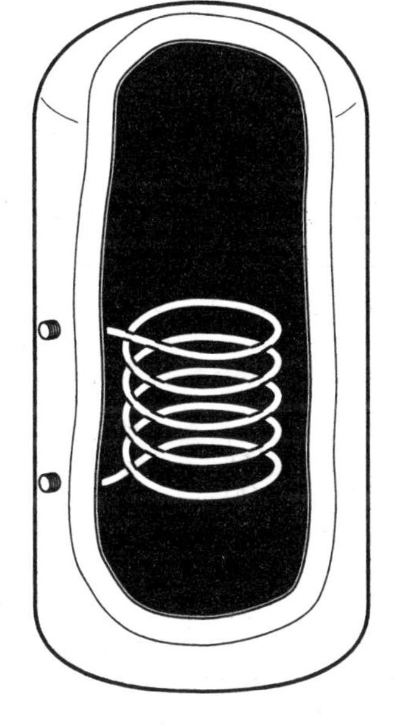

figure 2.5 a normal cylinder coil **b** high-performance cylinder coil

If you have an old style of boiler, it may be worth considering replacing it with a new one next time it needs any repair or maintenance work. This will reduce the time it takes to warm up the water and will in turn save money and provide better fuel efficiency.

Water temperature

The temperature of the hot water is set by the installer and should be adjusted to meet the needs of the end user. The temperature within a stored hot water cylinder should be adjusted to no higher than 60°C at the top of the cylinder. If it is set higher than this, the water may scald the user and limescale deposits may form in hard-water areas. Equally, the water should not be stored at a temperature much below this as the growth of legionella bacteria may occur.

Legionella

Legionella is rarely a problem in domestic homes. The bacteria are killed off above 60°C and will not survive very long at this temperature. However, they can survive within the temperature range of 20–45°C. They can be dangerous to humans and are transmitted when water is in a misty or vapour form, so areas around boosted shower outlet sprays could be vulnerable if the water is maintained at too low a temperature. The best alternative to use where cooler water temperatures are required is to store the water at 60°C and then use a blending/mixing valve, which mixes the hot water with a quantity of cold water to reduce the temperature to the desired level.

Direct systems of hot water supply

As the name suggests, direct systems include those in which the water is heated directly, such as by an immersion heater or by a boiler found some distance from the hot water storage cylinder. The heated water is transferred to the cylinder by gravity circulation (see below) via two pipes referred to as the primary flow and primary return. Where the water is heated in a boiler it is invariably quite hot and limescale build-up will occur inside the primary pipework in hard-water areas. Most direct systems are now quite antiquated and only the oldest of houses will still have such a system. The immersion heater, however, is still quite commonplace and makes an ideal backup when incorporated within the cylinder of an indirect system of hot water supply.

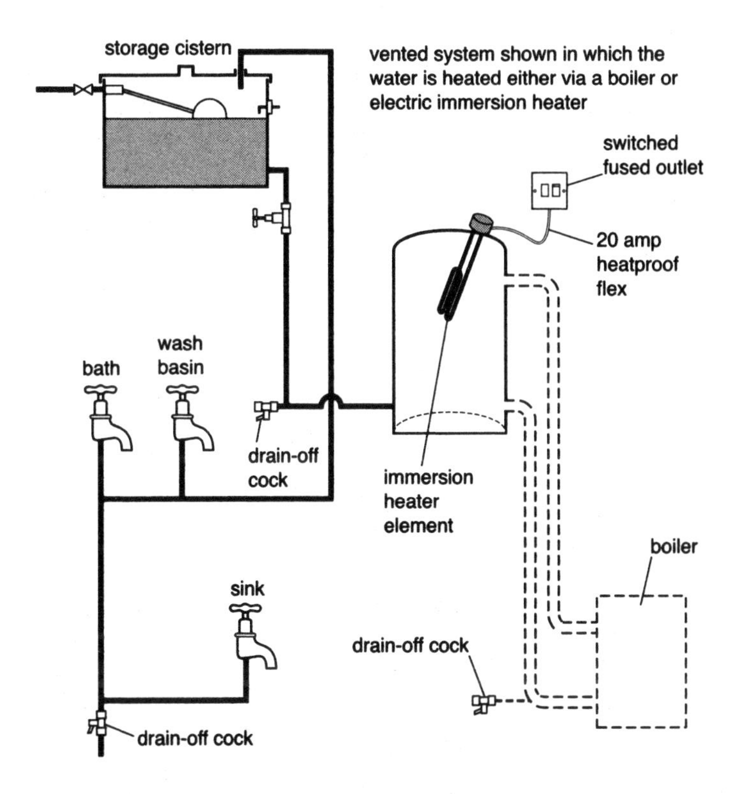

figure 2.6 a direct system of hot water supply

The immersion heater

This is effectively a large heating element like those found inside a kettle. When the immersion heater is switched on, the element heats up and remains on until the thermostat senses that the water temperature has reached its desired level or until the power is switched off. As mentioned earlier, the water should be stored no hotter than 60°C; this level is set by making an adjustment to the dial on the head of the thermostat. Where the immersion heater is fitted within an unvented hot water cylinder it will also require a high-limit cut-out thermostat set to operate (cut out) at 90°C. All new and replacement immersion heaters will include, as standard, this independent non self-resetting over-temperature safety cut out device to prevent the water in the cylinder from overheating.

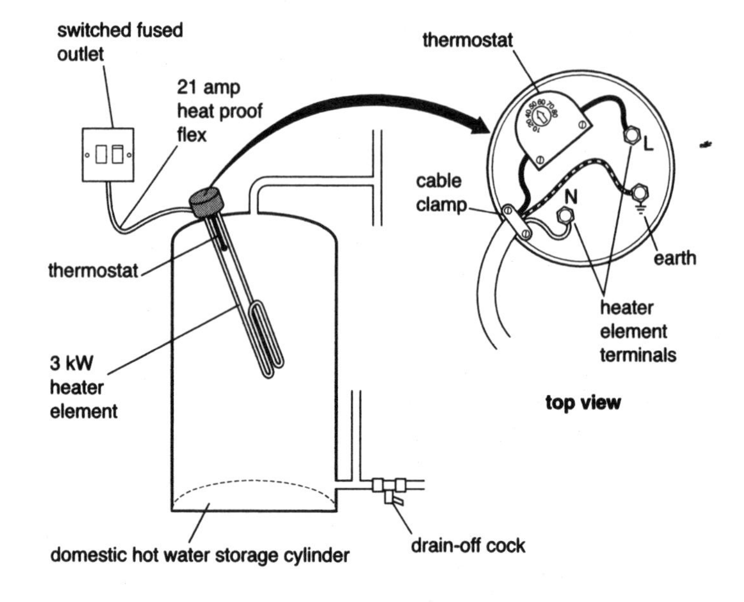

figure 2.7 the immersion heater

Gravity circulation

The hot water from the boiler in figure 2.8 is transferred to the cylinder by natural gravity circulation, i.e. hot water rises up the primary flow and is displaced by the column of descending cooler water within the primary return. This system is very common and will be found in a large number of properties. However, these systems are slow and the water in the cylinder can take anything up to two hours to heat up. Modern systems use a circulating pump to push this water around the circuit rapidly, allowing heat-up times of around 30 minutes, or sometimes even faster (see Chapter 03 for examples of fully pumped central heating systems).

Indirect systems of hot water supply

If you have a hot water cylinder in your home, there is a good chance it is part of an indirect system. With this type of system there are no problems with hard water scaling up the pipes, and central heating water can also be taken from the water heated within the boiler.

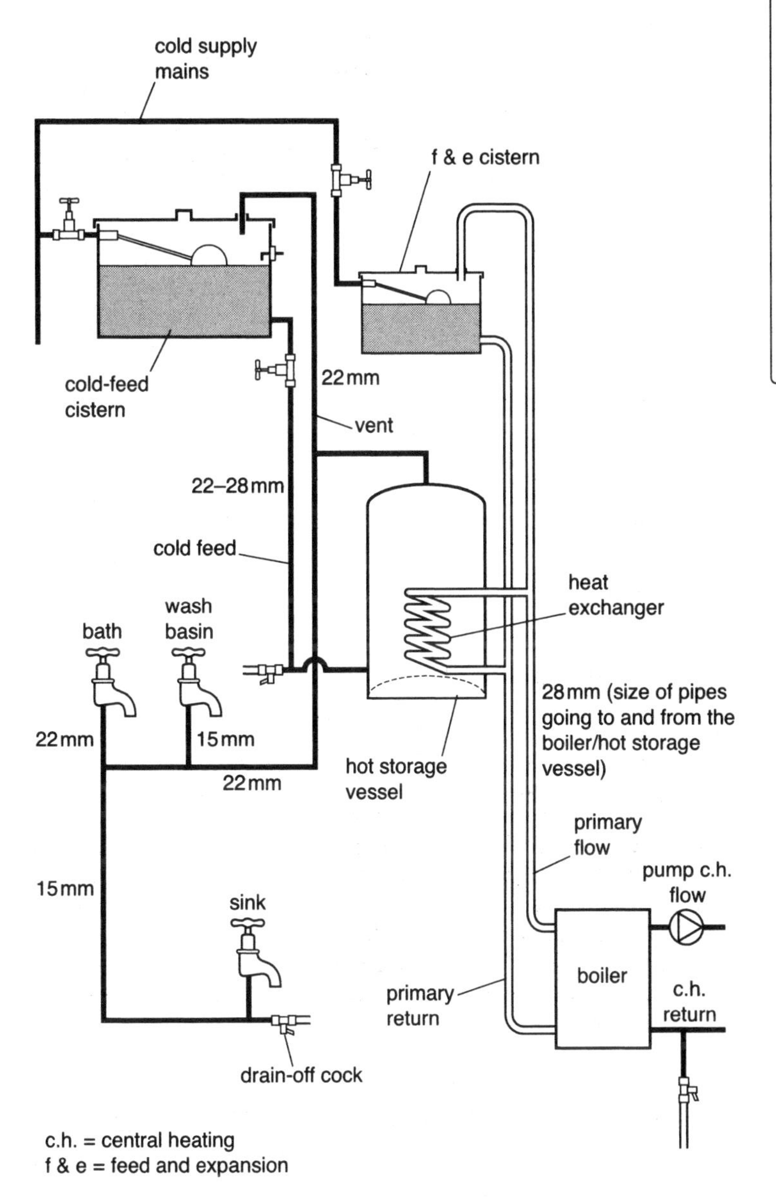

figure 2.8 an indirect system of hot water supply

Indirect systems of hot water supply have a heat exchanger coil located inside the hot water cylinder. This is, in effect, a pipe run in a series of loops inside the cylinder of water. Hot water from a boiler is passed through this pipe and the hot water flowing through the pipe coil in turn heats up the water in the cylinder. Thus the water is heated directly within the boiler, as in the direct system (referred to as the primary hot water) and indirectly via the pipe coil within the cylinder (referred to as the secondary hot water or domestic hot water (dhw)).

Indirect systems may be either vented or unvented. Vented systems are those in which the cold water is taken from a cold water feed cistern, usually found in the roof space; unvented systems are fed with cold water directly from the cold supply mains pipe. As can be seen in the vented system in figure 2.8, there are two separate cisterns within the roof space or loft. One is the cold water feed cistern, designed to supply water to the cylinder, and the other is a feed and expansion (f & e) cistern. An unvented system can be seen in figure 2.10.

Vented systems

Feed and expansion cistern (f & e cistern)

The f & e cistern ensures that the water contained in the boiler and heating system, where applicable, does not mix with the water used for the domestic hot water. There are two specific reasons for this separation:

- to combat the problem of limescale build-up
- to reduce the amount of atmospheric corrosion.

The water in the domestic hot water pipework is constantly being passed through the system, and with this will come a constant flow of oxygenated water containing a quantity of dissolved limescale. Figure 2.8 illustrates that the water which enters the boiler and heating system via the f & e cistern (which is heated to far in excess of 60°C) is never emptied unless it is drained out for maintenance purposes. So, limescale build-up is eliminated because once the water has been heated, no more limescale will be generated.

Also, after a short period of heating the water and moving it around the system with a circulating pump, all of the air will be expelled from the system. It is this air, in particular the oxygen in it, that causes the corrosion of iron, from which the boiler and radiators are invariably made, so losing this air prevents them from rusting. Corrosion is looked at in more depth in Chapter 06.

The water level within the f & e cistern

The water level within the f & e cistern is adjusted low down inside this cistern, just above the outlet. As the water within the system heats up it expands, rises back up the cold-feed pipe and is taken up into this cistern. If, during the installation of these cisterns, the water level is adjusted too high, the water, when heated, will expand and rise to a point where it will drip from the overflow pipe. Upon cooling, more water will re-enter the cistern via the float-operated valve and the process of overflowing will be repeated continually. This will result in fresh oxygenated and calcium-laden (limescale-forming) water continually being added to the system.

The open vent

You may be wondering why a pipe with an open end terminates above the water level within the cistern. Why is the vent pipe needed? First, it allows air in and out of the system during filling or draining down. You will notice that in figure 2.8 the water enters low down in the cylinder, near the bottom, and the hot water is drawn off from the top. If there was no vent pipe there would be a very large air pocket above the water which would prevent the water from filling the system. Also, when draining out the water from the system the vent pipe allows air to enter, which helps to remove the water.

The second purpose of the vent pipe is as a safety measure, ensuring that the system always remains at a pressure compatible with that of the atmosphere and allowing any pressure generated within the system to escape. A build-up of pressure could result from the cold feed to the system being blocked, as might happen if it freezes in winter or if debris accumulates inside the base of the storage vessel.

Hot water distribution

If you look again at the example of stored hot water supply (figure 2.8), you will notice that the hot water is drawn off from the top of the cylinder. The reason for designing the pipework in this way and taking the water from the top of the cylinder is that this is where the hottest water is found, because hot water naturally rises to the highest point. So, the cold water flows in at the base of the cylinder and pushes the hot water out when a tap is opened. If the cold water was supplied to the top of the cylinder it would mix with the hot water and cool it down.

Some cylinders are designed with the cold pipe entering at the top, which seems to contradict this argument but, in fact, if you could see inside the cylinder you would notice that the pipe extends down to the base of the vessel (an example of a dipped cold feed, as it is called, can be seen in figure 2.16).

Water expansion

When water is heated it expands by approximately 4 per cent from cold to 100°C. Above 100°C, at atmospheric pressure, it changes to steam and its volume immediately expands 1600 times. For safety reasons this expansion must be allowed for in the design of the storage cistern.

If you have an open-vented system, it will be under the influence of atmospheric pressure and as the water slowly heats up it will expand and be pushed back up through the cold-feed pipe into the cold-feed cistern that supplies the system. As mentioned above, if the cold feed becomes blocked, the expanding water will be forced to travel up the open vent pipe and discharge into the cistern, thereby preventing a pressure build-up within the system.

Imagine the possible danger if both the cold feed and vent pipe became frozen up and blocked. If the water were to heat up and expand, this expansion could not be accommodated and, as a result, the cylinder might split at the seams or even explode, hence the need to ensure that pipework is suitably insulated.

Single-feed system of indirect hot water supply

This type of hot water supply system is no longer installed today, but during the 1960s many of them were. They used a special indirect cylinder, which could fill the domestic hot water system as well as the boiler circuit with water, the latter of which also served a limited number of radiators. The design of this system falls outside the scope of this book, but it is mentioned here because, as there is no separate f & e cistern in the roof space, without this information you might think, when faced with such a system, that it is a direct hot water system. The primary water and secondary domestic hot water are separated by a trapped air pocket within the specially designed hot water heat exchanger.

The clue to knowing if you have this system in your home is to look on the side of the cylinder for the 'Primatic' brand name. Also, the single cold water storage cistern found in the loft with this system will serve several steel panel radiators, which it will not do if a direct system of domestic hot water is being used.

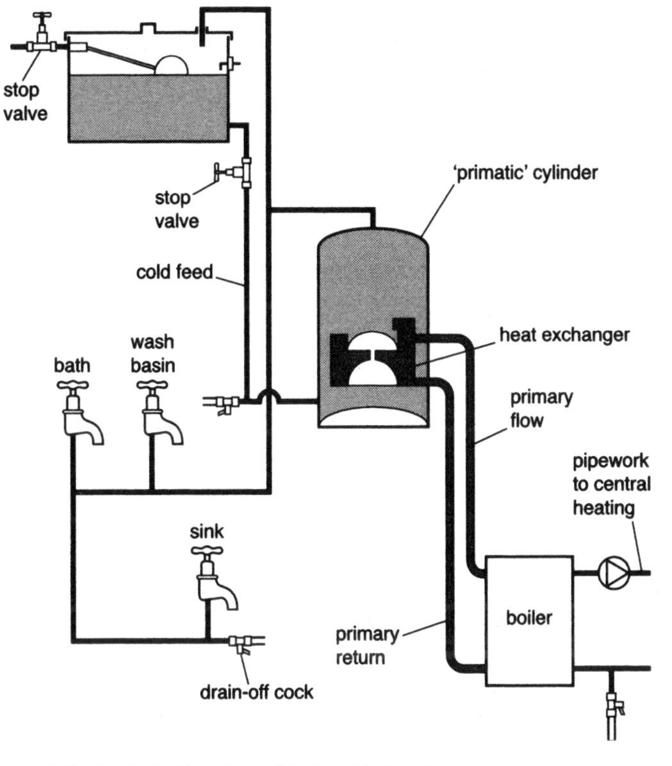

figure 2.9 single-feed system of indirect hot water supply

Unvented systems of hot water supply

Many homes built today incorporate this design of hot water supply. It has the advantage of:

- having a stored supply of potable hot water
- maintaining a good flow rate to the various outlet points
- being at water supply mains pressure
- freeing up the roof space to assist in the design of modern roof structures.

This type of system has only been permitted since 1985 and, as a result, is generally only found in newer developments or houses that have been refurbished. It is essential to note that the minimum size of the supply pipe to these systems is 22 mm – if it is any smaller you will not get the flow rate expected as compared to that of the vented system with its increased pipe sizes. New homes are constructed with this larger mains supply pipe, thereby generally posing no problems; existing properties, however, may only have a 15 mm inlet cold water mains supply and this will be inadequate to serve all of the hot and cold outlets within the property.

The installation of these systems falls within the requirements of the Building Regulations, as administered by the local authority, so the installation and maintenance of these systems must be undertaken only by approved operatives. This means that the installer will have taken and passed an assessment course aimed specifically at the design and safety of these systems.

Looking at the two systems shown in figure 2.10 you will see that there are several controls in addition to those found on the more traditional systems (as seen in figures 2.6 and 2.8). Two systems have been illustrated because one design uses a sealed expansion vessel to take up the expanding water whereas the other uses an air pocket, located inside the cylinder with a floating baffle.

The following notes identify a brief outline of the controls found on an unvented system, purely as interest and identification but, as stated above, remember it is a requirement that these systems are only installed and serviced by qualified personnel. Should you have such a system and require work to be completed on these controls, remember and ask to see the operatives approval certificate or card otherwise your house insurance may not be valid should something go wrong, as these systems can explode if not looked after properly.

Components of the unvented system

Strainer

This is designed to ensure that no grit or dirt within the pipeline can travel along the pipe and cause the ineffective operation of a control installed further downstream.

Pressure-reducing valve

This is a special control that prevents excessive mains pressures from entering the hot water storage vessel. The hot water storage vessels themselves are quite robust but will not withstand the highest of water pressures sometimes experienced within the mains supply. This control usually restricts the pressure to a maximum of 3 bar. In order to ensure equal

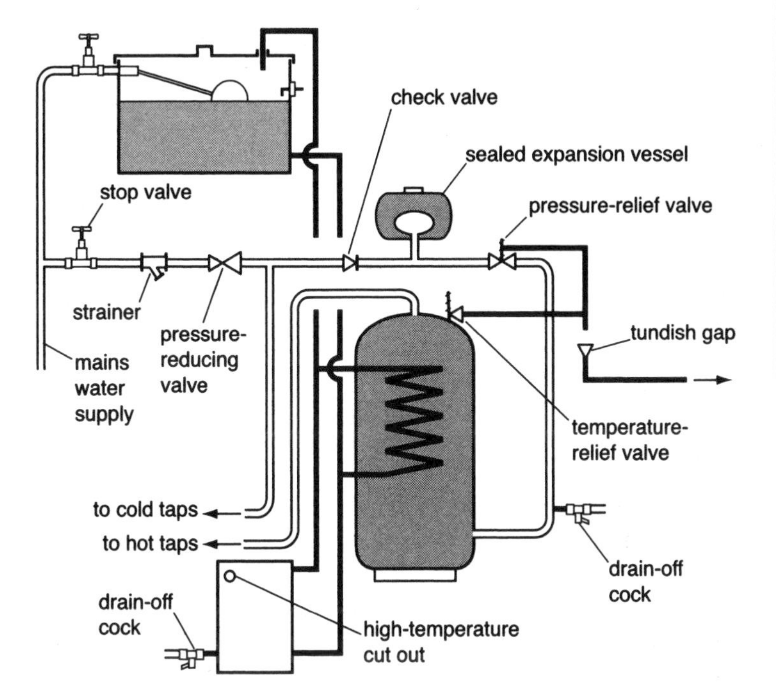

a) **system using a sealed expansion vessel (showing the water heated within a boiler)**

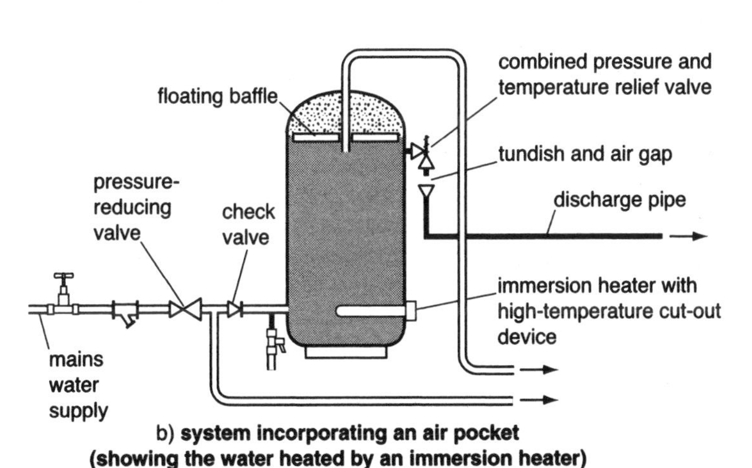

b) **system incorporating an air pocket (showing the water heated by an immersion heater)**

figure 2.10 unvented systems of domestic hot water supply

pressures in both hot and cold supplies, such as where mixer taps are incorporated, the cold water is sometimes branched off after this control valve, as seen in figure 2.10. Alternatively, a second pressure-reducing valve will be required on the cold-supply pipework.

Check valve

This valve is basically a non-return valve that has been incorporated to prevent the heated water expanding back along the pipework. It is a Water Regulations requirement that no water is allowed to flow in a direction opposite to that intended.

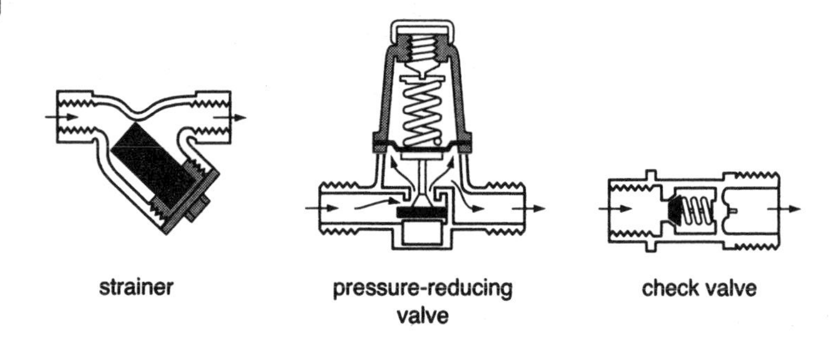

figure 2.11 components of the unvented system

Sealed expansion vessel

This unit is designed to take up the expanding water within a large rubber bag contained within an airtight vessel. As water is heated it expands and flows into the bag. This causes the air surrounding the bag to become pressurized; when the water cools, the air pressure forces the water back out into the system. Note that some systems do not use the sealed expansion vessel as identified here but take up the expanding water within an air pocket located inside the top of the cylinder.

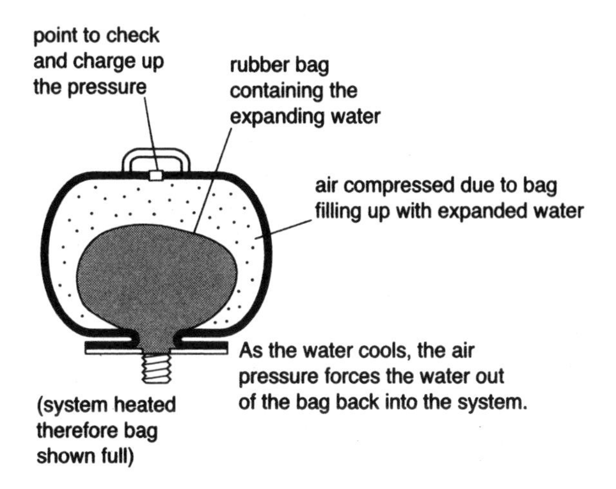

figure 2.12 sealed expansion vessel

High-temperature cut-out thermostat

This is basically a second thermostat in addition to the normal thermostat. This control will turn off the supply where the temperature within the system rises to 90°C. Should this control be activated you will need to manually reset the device.

Pressure-relief valve

This is a special control valve designed to open, allowing water to discharge from the system into a drain, should the pressure rise to such a point where damage to the storage vessel might result.

Temperature-relief valve

This is another special control valve designed to open should the high-temperature cut-out device fail to work. It allows the water to discharge from the system safely into a drain should the temperature rise to around 95°C, at which point it would become dangerous. With the high pressures that might be generated within the system due to heating water, the boiling point is increased and if the temperature were to get any hotter than this, uncontrollable steam could discharge from this control rather than controllable water.

Sometimes the pressure-relief and temperature-relief valves are incorporated within the same control valve and in both cases any water discharging from them is conveyed to the drain via an air gap and funnelled tundish. The air gap is maintained to ensure the drain pipework cannot make contact with the potable hot water supply pipework.

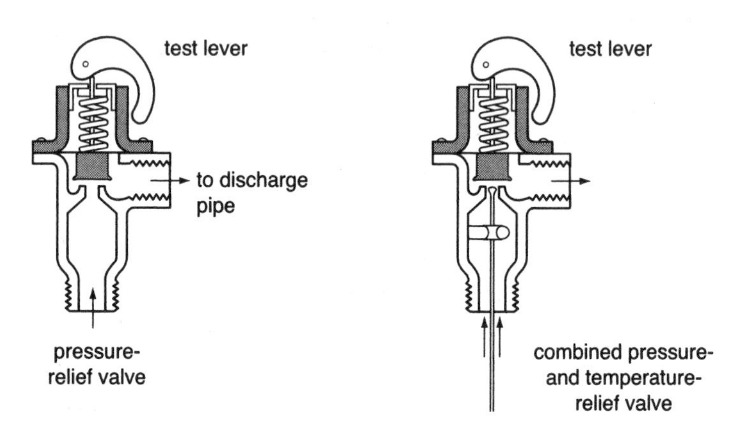

figure 2.13 pressure- and temperature-relief valves

Instantaneous systems of domestic hot water supply

The storage systems discussed above work well and a good flow rate of water from the taps can be expected from a correctly sized system. However, in the case of unvented systems for homes with many occupants or older properties with a small inlet supply pipe, which might be just 15 mm (half inch) in diameter, an instantaneous system may be the only choice where a connection to the cold mains supply pipe is made and has very much been the traditional system of domestic hot water supply.

Older properties that do not have heating systems often have an instantaneous system of hot water supply. They may have a centrally installed multipoint water heater or several single-point water heaters found at the appliances where the water is required. These heaters may be electrically operated or may be fuelled by gas.

Many homes have upgraded from the multipoint system by installing a combination boiler (often called a combi boiler for short) to supply hot water and central heating. These units heat water as it is required, rather than storing it at high temperatures, and also provide hot water that can be used for heating purposes.

The biggest drawback with the instantaneous water heater is the fact that the water can only be heated so fast and, as a result, the flow rate from the outlet tap is invariably slower than that expected from a storage system. The layout of the pipework to the various appliances is, however, the same as shown in figure 2.14 below.

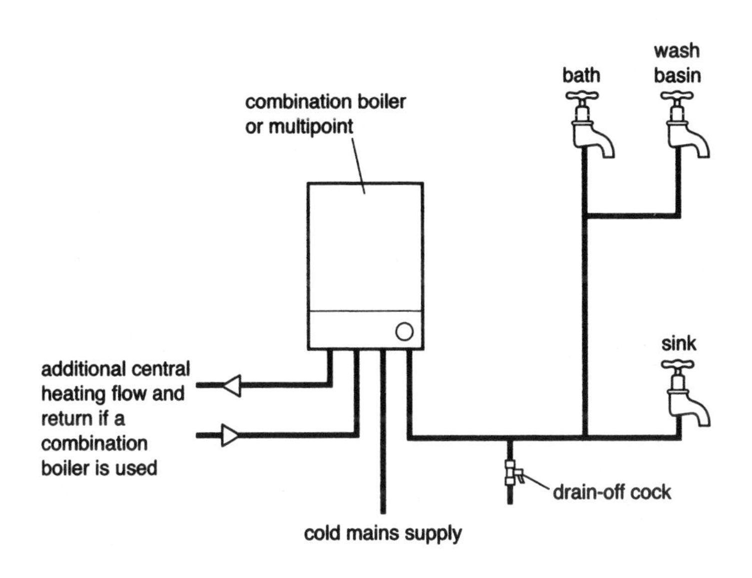

figure 2.14 centralized system of instantaneous domestic hot water using a combination boiler or multipoint

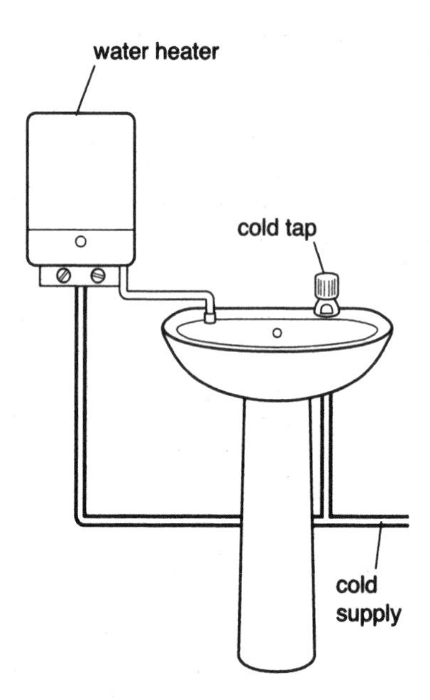

figure 2.15 localized single-point instantaneous hot water heater at the point of use

Thermal storage systems of hot water supply

These systems were introduced around 1985 as an alternative to the unvented storage system, having a supply of hot water at mains pressure, without all the necessary safety controls required for unvented systems. They are, in effect, a system of instantaneous hot water supply, taking their water directly from the mains supply. The difference between this and the unvented system is that this does not store hot water used for domestic draw-off purposes. Unvented systems are classified as such due to the fact that they contain a stored volume of water in excess of 15 litres.

If you look closely at figure 2.16 you will see that the storage cylinder is full of hot water, but it is not used for a domestic supply to the taps, as with all the storage systems previously identified, but only used to supply the heating circuit and thereby used to warm the radiators.

In the hot water container there is a pipe coil heat exchanger with many loops. If a hot water tap is turned on, the water will flow directly from the water mains through this coil, which causes the water to heat up rapidly, taking its heat from the cylinder full of hot water. It then passes through a blending/mixing valve which allows a percentage of cold water to mix with it if necessary, thereby cooling it to the desired temperature, as it may have become too hot when passing through the cylinder heat exchanger.

This system is by far the least common, but it is found in some homes. As with all systems that take their supply directly from the mains supply, it is essential that a sufficiently large water mains supply is available to prevent water flow problems. There is a similar system that contains the hot storage cylinder within the boiler case as one big unit, referred to as a combined primary storage unit (CPSU), which is essentially a variation of this design.

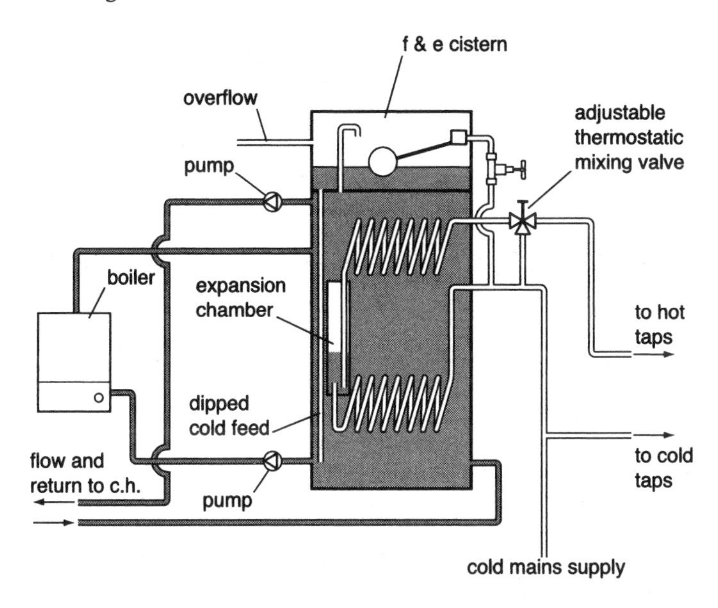

figure 2.16 thermal storage system of domestic hot water supply

Hot distribution pipework

Whether the centralized domestic hot water system is of the storage or instantaneous type, the water must flow around the building in pipes of appropriate size, reducing down in size to the smaller pipes that serve the various outlet points. A pipe of minimum diameter 22 mm needs to be used to supply a bath. As with the cold water pipework, a drain-off cock is located at the lowest point of the hot water pipework in order to facilitate draining down if necessary.

Choice of domestic hot water supply

What is best, a combination boiler or a regular boiler with a storage cylinder? This is a question that you will ask yourself when considering a new hot water supply. Each system has its own merits, and when designing a system you should weigh up the pros and cons in order to choose what is best for you. Some of the merits and pitfalls of each system are discussed below.

Combination (combi) boiler

A combination boiler heats up water for domestic use, providing hot water for the taps and for the central heating system. The installation of combination boilers currently makes up 60 per cent of market sales and therefore deserves the first consideration. However, it will not always be the best choice. The clear advantages include the following:

- easily installed and is the cheaper option
- only heats the water as and when it is required
- does not require a storage cylinder or cistern in the roof space
- water fresh from the mains is used for the hot supply to the taps
- the water will be at a good pressure for showers
- provides water for central heating.

There are lots of good points here, but this system also has disadvantages that are often overlooked when considering the installation of a new system of hot water supply. These include:

- a poor flow rate from the taps where the pipe size to the house is inadequate
- no boiler operation for central heating purposes when it is being used to heat the water for domestic hot water
- there will be no backup supply of hot water if the power or water supply is turned off.

Let's look more closely at these disadvantages. First, consider that the pipe entering the property is only 15 mm in diameter – you just might be expecting too much from the pipe. The modern home has dishwashers, washing machines, outside taps, numerous toilets and bathrooms. You cannot possibly expect this one pipe to feed all of these outlet points at once. It is unlikely that they would all be in operation at the same time, but several may well be and therefore something will be starved of water and the flow rate will drop dramatically. For two people living together this size may just be adequate, but where there are more people living in the home then this system should not be selected unless you are prepared to put up with the problems of poor flow, bearing in mind the boiler may not even operate if the flow rate drops below a certain level, as many require a minimum flow of water passing through the boiler.

Second, a combination boiler is a priority system, which means that when it is providing the hot water to the hot taps and other outlets it does not supply the heating system. In other words, the boiler gives priority to the domestic hot water when in operation; it does not do both heating and hot water at the same time. So, for example, in a home with say six people, every time the bath or shower is being run, or the washing machine requires hot water, or any hot tap is opened, the heating will not be on. As a result, you may find the radiators getting cooler on occasions.

Finally, the flow rate of water from the taps is less than that of a storage system. Systems fed from a storage cylinder seem to gush water through the taps when compared with the instantaneous systems, which need a little time to heat the water as it flows through the heater. Some combination boilers with very high heat outputs have combated this problem to a certain degree, but it must be understood that the bigger the boiler output the bigger the gas supply pipe, if this is the fuel used to feed the boiler. Bearing this in mind, is the gas supply pipe feeding your house sufficiently large? Some of the larger combination boilers require a gas pipe of diameter 28–32 mm.

Regular boiler and hot water storage cylinder

The advantages of having a stored vented domestic hot water supply are generally the opposite of the problems of the combi boiler and include:

- The water flow out from the taps will be good (this is not to be confused with pressure, as previously identified).

- The central heating is independent of the hot water (i.e. this is not a priority system).
- The kilowatt rating or output size of the boiler does not need to be as great.
- There will still be a limited backup supply of hot water if the water mains supply is turned off.

The points above relate to a vented storage system. Note that if an unvented system is installed a large supply mains is still required to combat poor flow conditions (minimum 25 mm polyethylene). The disadvantages of the storage system are the opposite of the advantages of the combination boiler:

- More pipework is required for installation, therefore it is more expensive to install.
- Water is heated for domestic hot water purposes, even if not required, so it can be more expensive to run.
- Additional room is required for the storage cylinder and the cold water cistern.
- Where the cold storage cistern is not located high enough, very poor pressures will be experienced from outlet points, particularly shower outlets, therefore additional shower boosters may be required.

So, in conclusion, if there are only two or three people living in the property, the minimum pipe diameter is 22 mm and the occupants are prepared to wait a minute or two longer to run their bath, then a combination boiler might be a suitable system. Money will be saved on installation and on running costs.

However, where several occupants inhabit the home, creating a greater demand for hot water, it might be worth finding the space to incorporate a regular boiler and hot storage cylinder, preferably unvented, thereby ensuring good flow and pressure to all outlet points without disrupting the central heating demand. Of course, this is dependent on the mains supply pipe being big enough, otherwise a vented system of stored hot water should be used.

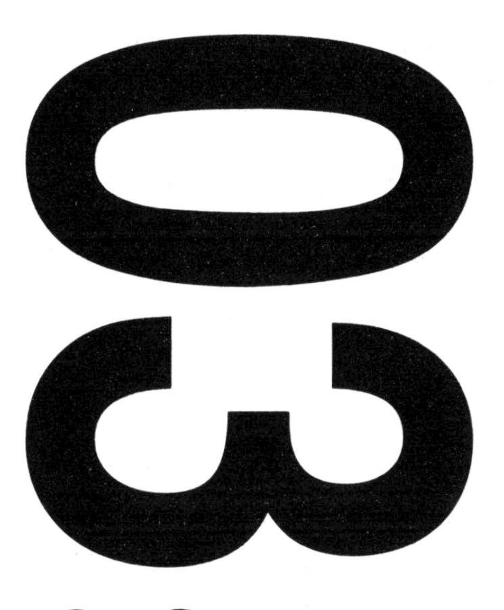

domestic central heating

In this chapter you will learn:

- **about the different types of central heating**
- **about central heating boilers**
- **about central heating controls**
- **how to protect heating systems.**

This chapter explains methods of central heating and the fuel used to keep you warm. There is a variety of different methods of domestic central heating, including:

- electric storage heaters
- warm-air heating
- under-floor heating (radiant heating)
- water-filled radiators.

Of the above, water-filled radiators is by far the most common system and therefore will be the main focus of this book. Of the others:

- electric storage heaters use cheap-rate electricity at night to warm up heat-retaining blocks, designed to slowly release their heat throughout the day
- warm-air heating consists of a series of ductwork to distribute pre-heated warm air around the home
- under-floor heating uses either heated electric cables or water-filled pipe coils to warm up the structure of the building.

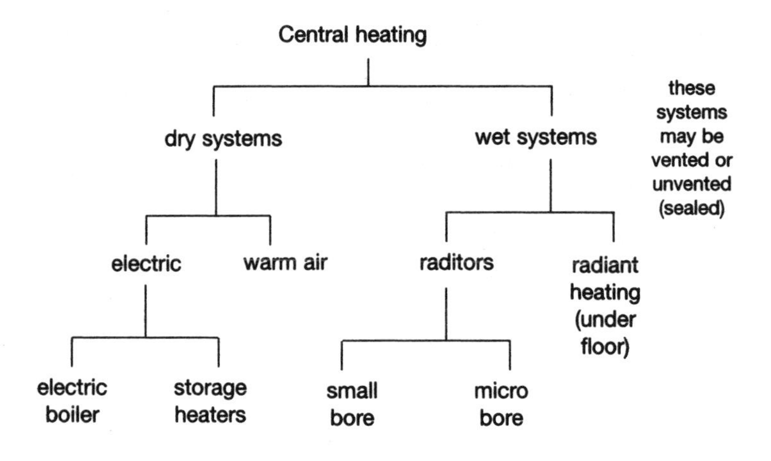

figure 3.1 types of central heating systems

More and more new developments are including under-floor heating as an alternative to the more traditional radiator system. Under-floor heating is referred to as radiant heating and merits a review of its design in order to understand how it works effectively compared with water-filled radiator systems.

Radiant heating

Radiant heating uses infrared heat rays that do not warm up the air through which they pass but the structure upon which they fall. In other words, radiant heating does not warm up the temperature of the air in a room; instead, it warms up the structure of the building.

When a person enters a room their body tries to become the same temperature as the surrounding structure and, as a consequence, if the building is cooler than you are, infrared heat is lost from your body as it tries to even out the temperature difference. If, however, the structure of the building is warm, no heat will be lost from your body in this way. As a consequence, the ambient temperature of the room can in fact be cooler than your body and the building as the air temperature does not unduly affect your body temperature.

Coils filled with water are laid within the floors and if they are left on long enough at a temperature of around 40°C, they will emit sufficient radiant heat to slowly warm up all the surfaces and solids within a room to a temperature compatible to that of the human body – around 33°C.

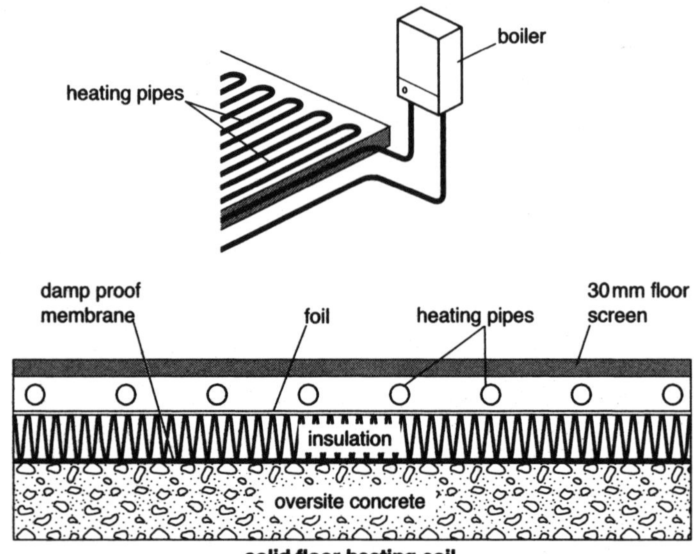

figure 3.2 radiant heating

The advantages of having a radiant heating system include:

- cooler room temperatures which create a sense of freshness
- less transference of dust and airborne bacteria caused by the effects of convection currents
- very low water temperatures resulting in greater efficiency from the boiler (typically around 90 per cent). Efficiency is explained later – see page 87.

Traditionally, UK homeowners only put on their heating for a few hours in the morning and a few hours in the evening. This limited amount of time is rarely sufficient to warm the whole building and, as a result, higher flow temperatures are used to warm the structure. This creates a certain amount of discomfort underfoot due to the elevated water temperature, and also reduces the efficiency of the boiler. For these systems to work really well, long periods of low-temperature water heating is required.

The other major disadvantage of this system is the problems created by a leak in the pipe coil. Fortunately leaks are quite rare, but it can prove very costly to find the leak and make the repair.

Central heating systems using radiators

Unlike under-floor heating, traditional water-filled radiators warm up the air surrounding the large metal surface of the radiator. It is this warming of the air that creates convection currents within the room. Convention currents are the flow of warm air around the room, caused by the hot air rising as it expands and becomes lighter, and the cooler, heavier air falling to replace the void – the cycle continues until the room is warm.

The pipe layout of this sort of central heating system can be of several designs, although around 95 per cent of all domestic heating systems using radiators use what is called the two-pipe system. The two-pipe system consists of two pipes leaving the boiler. These two pipes, the flow and the return as they are called, travel around the house to the various radiators. At each radiator a tee connection is made to a pipe that branches off to feed a valve, usually found at either end of the radiator. The two pipes terminate at the last radiator.

For the last 50 years or so a circulating pump has been used to circulate the water around the heating system. Very rarely, in older properties, gravity circulation systems can still be found (see Chapter 02). Sometimes these systems use solid fuel (wood or coal) and, unlike gas- or oil-burning appliances, you cannot simply switch off the flame, so a radiator or two is incorporated as a heat leak from the boiler, allowing heat to escape naturally from the boiler by gravity circulation. However, these systems are now quite antiquated and ought generally to be replaced.

Other central heating designs, such as the one-pipe circuit or the reversed return system, can also be found, but due to their rarity in the domestic home they fall outside the scope of this book and have been omitted to avoid confusion. See **appendix 3: taking it further** for further reading on these systems.

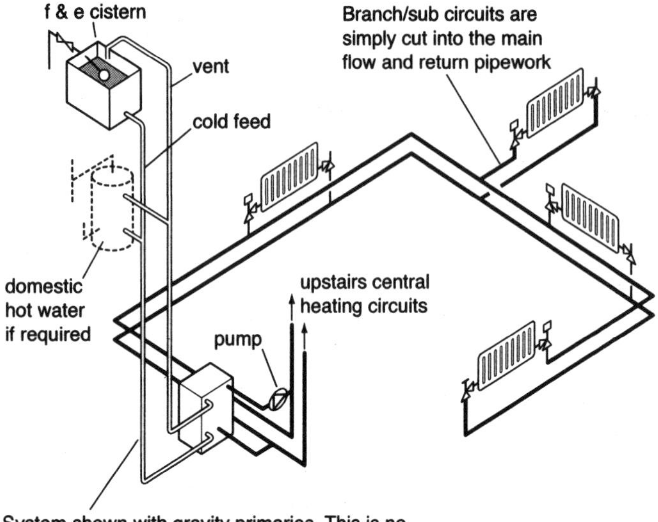

figure 3.3 two-pipe system of central heating

The water to the system shown in figure 3.3 is supplied via a feed and expansion (f & e) cistern found in the roof space (see Chapter 02). This type of design is referred to as a vented system; however, the water may have been fed directly from the cold supply mains via a special filling point, in which case the system would be referred to as a sealed heating system and as such would not be under the influence of atmospheric pressure.

Note also that the boiler is used to heat up the domestic hot water. In the system shown, a circulating pump is only used to force the water around the heating circuit. The water in the hot water cylinder circulates due to the effects of gravity (i.e. convection currents where the lighter hot water rises and heavier cold water sinks, as discussed in Chapter 02). This design does not comply with current Building Regulations but may, nevertheless, be the system that you have. Modern systems use a circulating pump to provide a more efficient system (a fully pumped system) as shown in figures 3.4 and 3.8.

The installation of a modern central heating system fuelled by either gas or oil must comply with the latest edition of the Building Regulations. Systems that were installed prior to the current laws do not need to be updated, but should you replace your boiler at some time in the future you will need to upgrade your system as appropriate.

Sealed heating systems (closed systems)

What is a sealed heating system?

The term sealed system relates to systems that, once they have been filled up, usually via a temporary cold mains connection, have the temporary hose connection removed and the system closed off. The water is now trapped within the system and as such it is not under the influence of atmospheric pressure.

Combination boilers are installed as sealed systems. They are designed with a temporary mains water filling connection to the central heating water and a permanent cold water mains supply for the domestic hot water draw off. The reason the temporary filling connection is disconnected from the water supply is that:

- it is a water regulation requirement
- chemicals may be added to the central heating pipework and if these are drawn back into water authorities' water supply it would lead to contamination.

It should be noted that the temporary hose connection must actually be disconnected from the supply in order to comply with the regulations and not left connected with the valve simply turned off as is often the case.

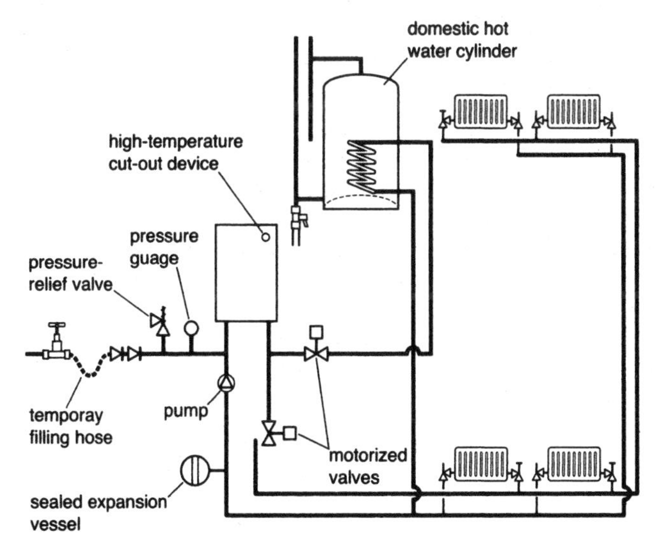

figure 3.4 sealed heating system

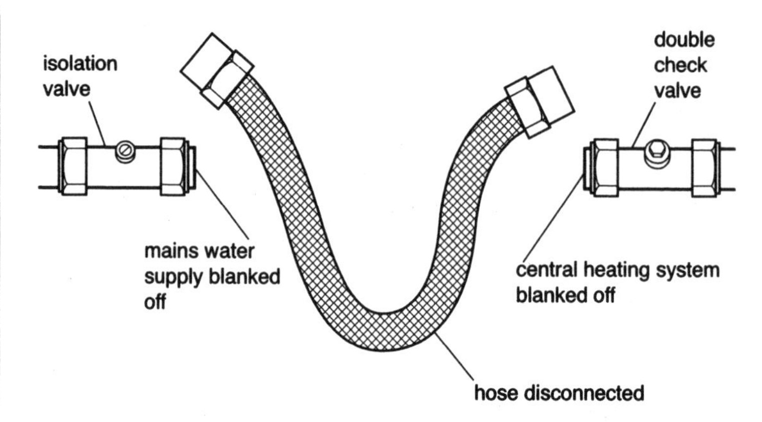

figure 3.5 temporary filling loop

How are sealed systems different from vented or open systems?

The water in these systems is trapped within a closed circuit and therefore is not subject to the influences of atmospheric pressure. The expansion of the water, due to it being heated, is accommodated within a sealed expansion vessel. This expanding water creates additional pressures within the system and cause it to rise in excess of 1 bar pressure; in fact, these systems are invariably slightly pressurized, as a manufacturer's requirement upon filling to a typical pressure of $1^1/_2$ bar. As the pressure increases within the system so does the temperature at which water boils. This could prove dangerous should excessive pressures develop, so the following safety controls need to be included at the time of installation:

- a temperature cut-off device, designed to shut down the appliance if the temperature exceeds 90°C
- a pressure-relief valve (safety valve) which can open to relieve the pressure from within the system if it becomes too great.

Sealed expansion vessel

In the case of vented systems of central heating, the water expansion resulting from the heating process is accommodated within the f & e cistern. Sealed systems, however, do not have this cistern open to the atmosphere, therefore the expanding water is taken up within a special steel container often found within the boiler casing itself.

The vessel contains a rubber diaphragm that separates it into two compartments. One side is filled with air to a pressure equal to that of the water in the system when it is cold; the other side the system allows water to flow in and out as necessary. As the water heats up it expands and enters the vessel, pressing against the diaphragm and squeezing the air on the other side of the diaphragm into a smaller space, thus causing the pressure to increase. When the system cools the increased air pressure forces the water back out into the system. Note that this expansion vessel is of a different design from that used for a system of unvented domestic hot water, where a rubber bag is used to contain the expanding water (see figure 2.12).

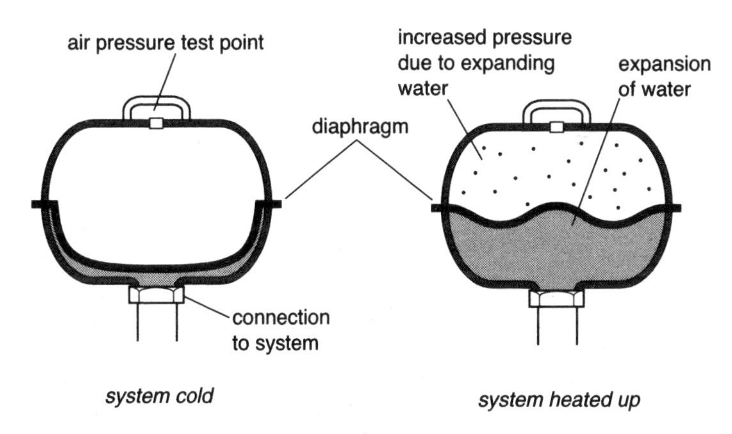

figure 3.6 operation of a sealed expansion vessel

Fully pumped systems and the location of the circulating pump

In a fully pumped system the circulating pump creates pressure within the pipework. It creates a positive, or pushing, force as the water is thrown forward from the pump and a negative, or sucking, force as it is drawn back into the pump when it returns from its journey around the system.

In the case of the sealed system, as shown in figure 3.4, often the pump is incorporated within the boiler, installed on the pipe as it leaves the boiler. Because a sealed system is not subject to atmospheric pressure, half the system is subject to positive pressure and half to negative pressure. The pressure gradually reduces from the pushing force to zero, and the suction slowly gets stronger as the water returns to the pump. As a consequence, provided that there are no leaks, air cannot be drawn into the system.

This is not the case with a open-vented system. The diagrams in figure 3.7 illustrate the principle that the cold feed enters the system at the point where the influence of the pump changes from positive to negative pressure. This point is referred to as the neutral point.

In figure 3.7(a) the system is working well – the pump is creating positive pressure around the whole system, which ensures that no micro-leaks (very small openings allowing the passage of air but not water) will allow air to be drawn into the system. In this same system, if the pump were installed the other way round it would create a negative influence throughout. This would work fine, but air could be drawn in through radiator valve gland nuts, where the spindle turns (a typical micro-leak). Therefore, to ensure a good design always aim to get a positive force.

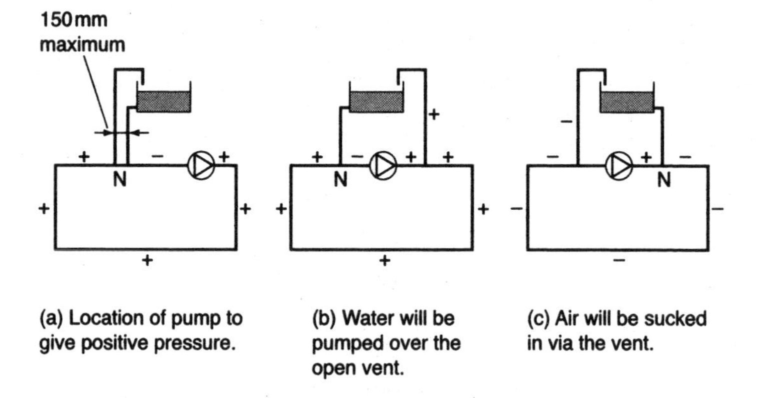
150 mm
maximum
N
N
N
(a) Location of pump to give positive pressure.
(b) Water will be pumped over the open vent.
(c) Air will be sucked in via the vent.

figure 3.7 the principles of correct pump location

However, in the system in figure 3.7(b) the open vent pipe is under a positive influence and therefore will allow a quantity of water to discharge into the f & e cistern, subject to the head pressure created by the pump, and in so doing will oxygenate the water.

In the system in figure 3.7(c) the open vent is subject to the negative pressure of the pump, so air will be drawn into the circulatory pipework. This configuration is often overlooked as it is not so easy to spot. It can be identified by submerging the open vent in a cup of water – if it is sucking air into the system it will suck the water up from the cup.

When air is being drawn into your installation, not only is this inconvenient, causing radiators to fill with air and preventing them from working correctly, but it is also slowly and surely corroding your system from within as the oxygen in the air, when mixed with water, causes the iron radiators rapidly to corrode and rust away. The key thing to check for with an open-vented fully pumped system is that the open vent connection is within 150 mm (6 inches) of where the cold feed joins the circulatory pipework.

The air separator

Sometimes heating installers incorporate an air separator into the pipework to serve as the collection point for the cold feed and open vent pipe. This fitting ensures the required close grouping of the cold feed and vent is maintained and also creates a situation where the water becomes shaken and turbulent as it flows through the fitting. This assists in trapping the air molecules within the water to dissipate and escape by forming and rising up out of the system through the open vent.

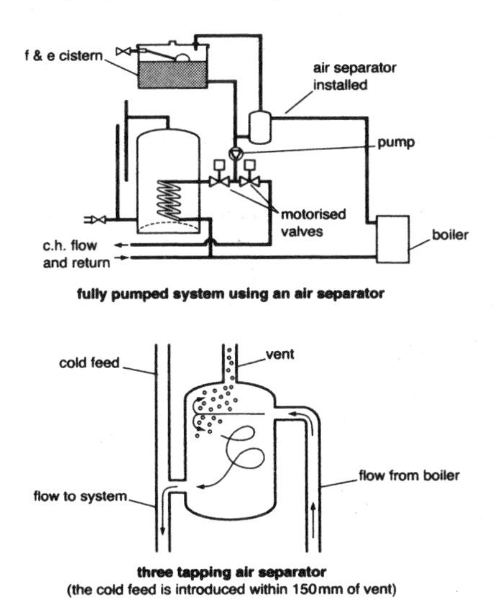

figure 3.8 use of an air separator

Micro-bore systems

Micro-bore is the name given to a central heating design that uses very small water pipe diameters. The pipe layout initially looks different but in fact it still follows the same design principles of the two-pipe system. Look carefully at the illustration of the micro-bore system in figure 3.9 and you will see that a flow and return connection has been run from the boiler to each radiator. The main difference between micro-bore systems and the usual systems using 22 mm and 15 mm pipework is that instead of using tee joints at the connection to each radiator a manifold is employed, from which several branch connections are made. In order to show another variation on the theme of central heating design, the micro-bore system shown in figure 3.9 has been run from a combination boiler.

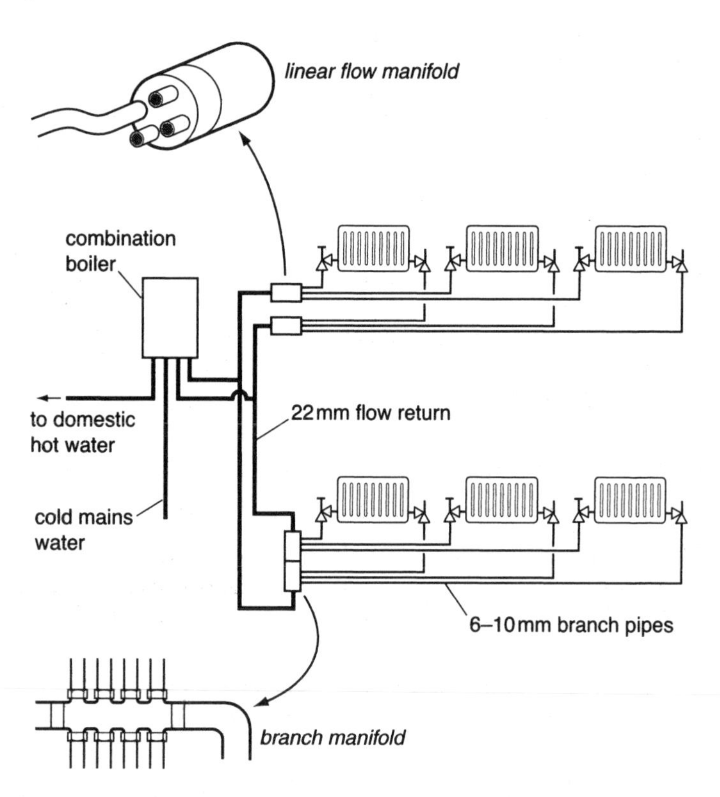

figure 3.9 micro-bore heating system

Radiators and heat emitters

There are many different types of radiators, including modern fancy shaped towel rails, skirting heaters, panel radiators, convector heaters, old-fashioned cast iron sectional column radiators – call them what you like, but they all basically do the same job of warming the room in which they are installed. They warm the air in close contact with the radiator, and convection currents circulate the warm air around the room, as discussed earlier. Some designs are more effective than others, for example the convector heater incorporates metal fins to assist in the distribution of the heat from the radiator.

Manufacturers indicate the heat distribution from a particular heater as its kilowatt output (i.e. the higher the kilowatt output the greater the heat that can be produced). This must be taken into consideration when fitting heaters as it would be useless to install a radiator that is too small for a room as the occupants would feel insufficiently warm; likewise, a radiator that was too large would occupy more wall space than necessary and would make the heating system far less efficient. The room may warm up more quickly, but the amount of fuel used to heat up the larger volume of water contained within the radiator would increase.

The size of heater for a particular room can be calculated; however, this requires the use of special tables and calculations pushing it beyond the scope of this book. The process is not, however, particularly complicated and those who are interested in learning more should see the further reading suggestions in **appendix 3: taking it further**.

Radiator valves

A control valve will be fitted to each end of your radiator:

- one is designed to open and close the radiator
- the other valve, referred to as a lockshield valve, is fitted at the other end of the radiator, is non-adjustable and will have a plastic dome-shaped cap.

The first valve, used to open and close the radiator, may consist of a simple plastic headed on/off control valve or a thermostatic radiator valve (TRV for short). For many years it has been the heating system installer who has chosen whether to use a manual valve or TRV, but current Building Regulations dictate the use of TRVs. The only radiators that can be fitted with a manual valve are those connected to radiators in rooms where a room thermostat has also been installed.

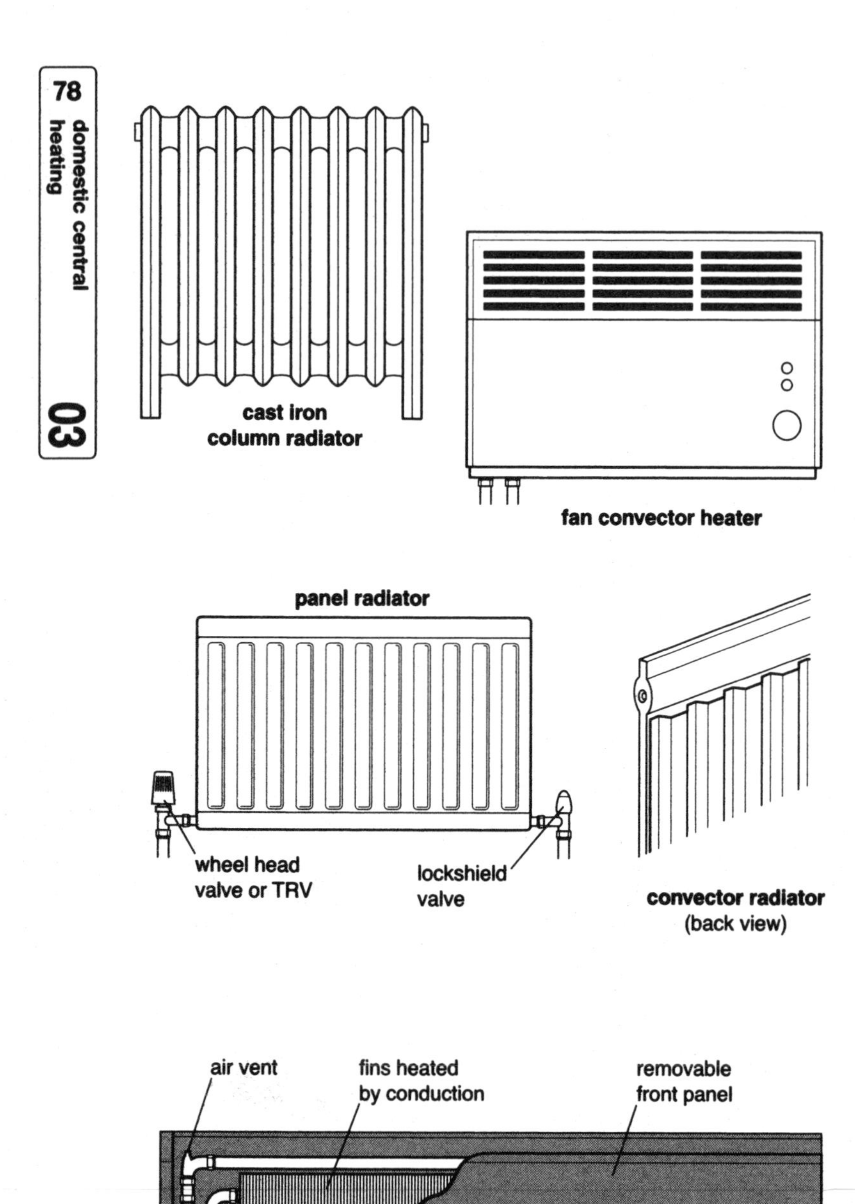

figure 3.10 types of radiators and heaters

The TRV automatically closes off the water supply to the radiator when the room has reached the desired temperature and therefore saves on the amount of fuel being used to continually supply heat wastefully to a sufficiently heated room.

The lockshield valve, located at the other end of the radiator, has a specific purpose in that it controls the amount of water that flows through the radiator. It is identical to the manual on/off valve except that the plastic head does not have an internal square socket to fit over the turning spindle of the valve. This valve has been pre-adjusted by the installer with a spanner at the time of installation when balancing the system.

Balancing

In order to ensure that the first radiator on the heating circuit does not take all of the hot water flow from the boiler, due to the water taking the shortest route through this first heater rather than going around the whole system, the lockshield valve is partially closed. By having this valve open by, say, only half a turn, the water is forced to continue along the heating circuit to the next radiator. Further radiator lockshield valves are also adjusted as required to force the flow of water throughout the whole system.

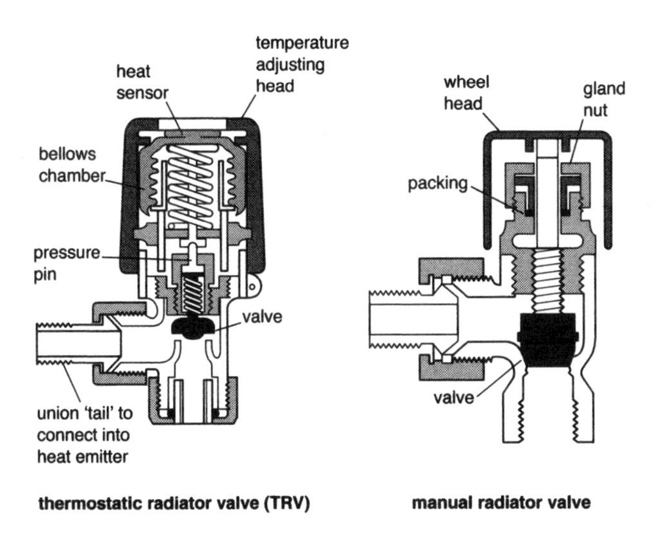

figure 3.11 radiator valves

Knowing which is the first and which is last radiator in the system

Basically, when you turn on the heat source of a cold system, the first radiator to get hot is the first or nearest to the boiler and will have the shortest circuit, the next to heat up will be the second radiator on the circuit and so on throughout the system.

Note: If you ever need to turn off the lockshield valve with a spanner, for example when removing the radiator for decorating purposes, remember to count the number of turns to close the valve, so that when you re-open this valve you open it by the same number of turns, otherwise you might find that some radiators on your system do not reach their desired temperature because you may have affected the balancing of the system.

You should also note that if you have a micro-bore system then sometimes both valves are found at one end. This is achieved by utilizing an internal tube in the radiator to distribute the water flow as necessary, as shown in figure 3.12.

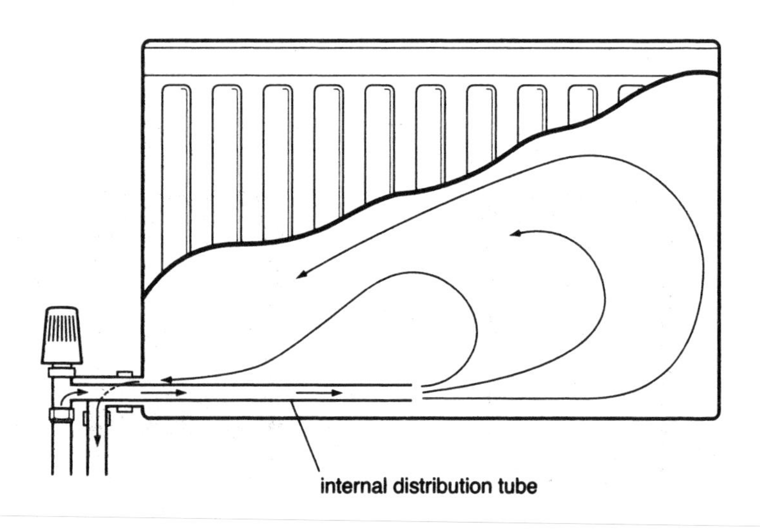

figure 3.12 micro-bore connections at one end of a radiator

Air within the system and air vents

Prior to water entering the central heating system air will be inside. As water enters the system of pipework the air will be trapped in high pockets and, as a consequence, will prevent the system from operating correctly. This air is expelled from any high points such as the tops of radiators by small openings into which air-release valves have been installed.

The installer of the system will aim to run the pipework in such a way as to avoid trapping air. Where this is unavoidable, an automatic air-release valve is inserted in the pipeline. This device contains a small float with a valve attached to its top end, so if water is present the float rises and the valve blocks up the outlet, whereas if there were no water within, the float would drop and open its outlet point or vent hole.

In addition to letting air out of the pipework and radiators, it is also necessary to open any air-release points when the system is being drained down, otherwise it will take forever to empty as air needs to enter the system in order to facilitate the removal of the water.

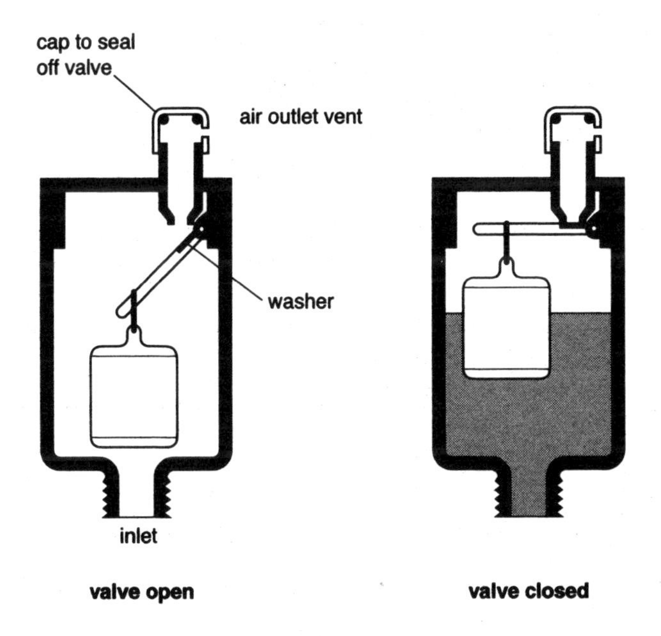

figure 3.13 automatic air-release valve

The boiler

What about the heat source for the system? In a nutshell, in its most fundamental form this is simply a metal box that is surrounded by a fire. In fact, the first heating systems were just this, a metal box referred to as a back boiler, found within the fireplace of the lounge. Surprisingly, there are a few still kicking around out there in some of the older properties.

Boilers today are fully automatic devices that turn up the heat as necessary and, with the exception of solid fuel systems, completely turn off when not required. The word boiler is used but in fact the water never actually boils within the appliance, or in any case it should not, otherwise there would be something drastically wrong. The water is just heated until the required temperature is achieved, as set by its built-in thermostat, and then the heat source turns off. The fuels that could be used for the boiler include:

- solid fuel, including coal, wood and straw
- electricity
- gas
- oil.

Electric boilers are quite rare and as such fall beyond the scope of this book. The remaining fuel types, however, have been used for boiler design for many years, and the design of the boiler has developed into a very efficient appliance, unlike that of yesteryear.

Solid fuel has limitations in its design and because these boilers tend to be more labour intensive, i.e. you need to load the fuel and empty the ash, they clearly are not as popular and account for around only 0.5 per cent of all installations. Around 92% of installations use gas and the rest use oil.

Due to developments over the years, many boiler designs and many different manufacturers can be found, with a never-ending list of models applicable to a particular design. But fundamentally they all fall into one of four basic types:

- natural draught open-flued
- forced draught open-flued (fan assisted)
- natural draught room-sealed
- forced draught room-sealed (fan assisted).

So what do all of these names mean? Essentially, they relate to the method by which air is supplied to the boiler.

- Natural draught or forced draught indicates whether or not the appliance has a fan incorporated to assist in the removal of the combustion products to the outside.
- Open-flued boilers take their air from within the room where the boiler is found.
- Room-sealed means that the air is taken from outside the building.

Referral to figure 3.14 provides an illustration of these four designs.

The boiler in your home will be of one of these designs. For example:

- If you have a back boiler located behind a gas fire in the lounge, you have a natural draught open-flued boiler.
- If you have a large freestanding boiler in your kitchen, with a flue pipe coming from the top, travelling into a chimney or passing through a pipe to discharge up above the roof, again this is likely to be a natural draught open-flued boiler.

Both of these boilers take their air from the room in which they are installed and this air is replaced via an air vent to the outside.

If your boiler has a terminal fitting flush with the wall it is most likely to be a room-sealed appliance.

- If this terminal is quite large it will be of the natural draught type.
- If it is small, say about 100 mm in diameter, it will be fan assisted.

These boilers do not take the air required for the combustion process from the room, but directly from outside.

There are many variations of boiler design, dictating the location of the fan or the route of the flue pipe (which may be vertically through a roof or horizontally out through the wall), but they all fall within the four basic types listed above.

In addition to the basic boiler designs, boilers are further classified into four generic types:

- non-condensing regular boiler
- non-condensing combination boiler

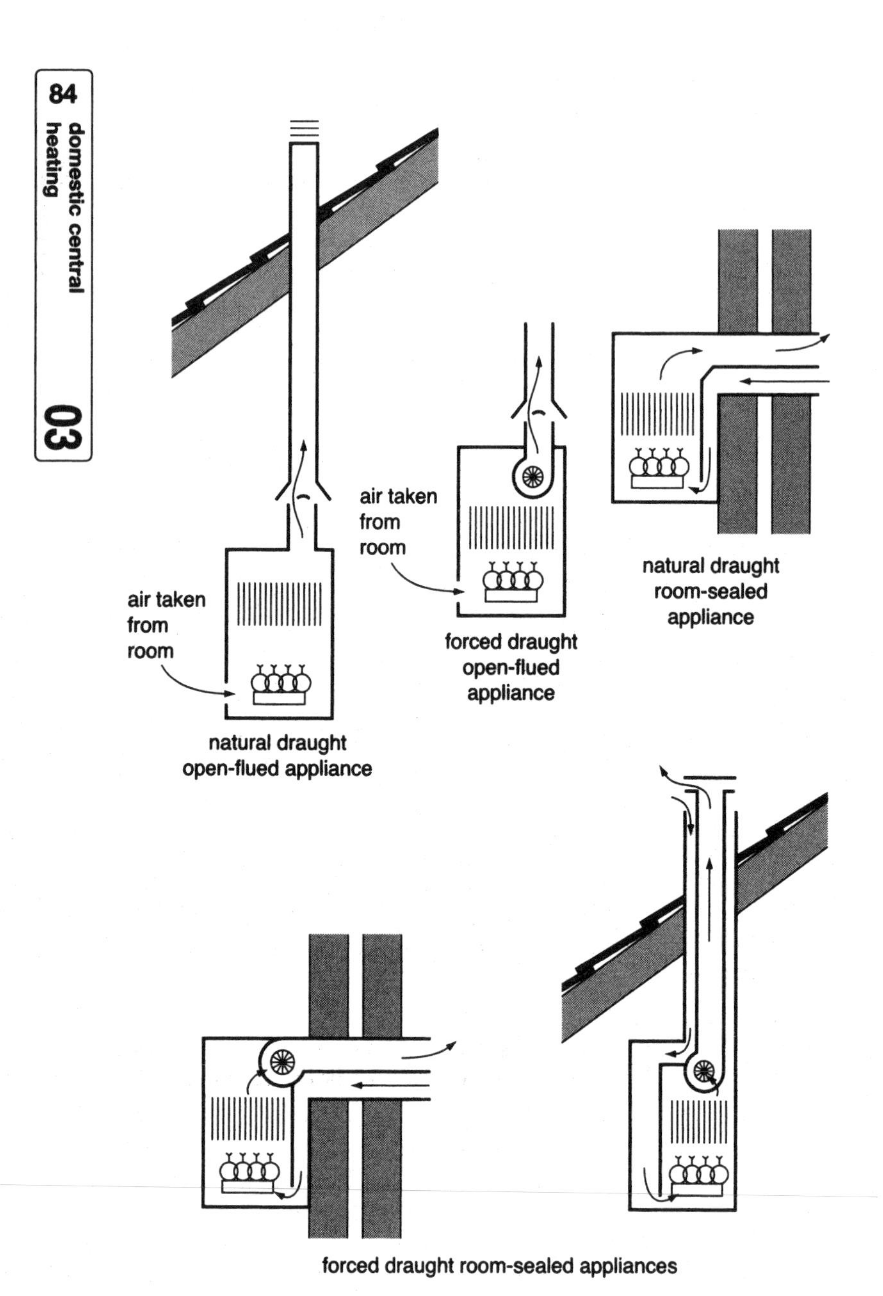

figure 3.14 boiler designs

- condensing regular boiler
- condensing combination boiler.

The differences between regular boilers and combination boilers have already been discussed in Chapter 02, but a new term is use here: 'condensing'.

What is a condensing boiler?

This is a boiler designed to take as much heat from the fuel and combustion products as possible and, as a result, is much more efficient. It is sometimes referred to as a high-efficiency boiler.

All domestic boilers installed prior to 1988 were designed in such a way that no consideration was given to the heat contained within the combustion products that were discharged from the boiler. If you were to take a thermometer and measure the temperature of the flue gases as they left the terminal, you would get a reading of something like 160°C. This is clearly a waste of heat and therefore of fuel. The condensing boiler is designed in such a way that these combustion products are cooled to as low a temperature as possible, using their heat energy in the process.

For the traditional central heating system using radiators, this flue temperature would be somewhere around 80°C. This temperature could be reduced even further to, say, 45–50°C where a radiant heating system was installed (as discussed earlier). Where the appliance reduces the flue products down to a temperature of less than 54°C (i.e. the dew point of water) water that is produced as part of the combustion process condenses and collects within the boiler and is subsequently drained from the appliance.

These boilers, when in operation, especially when it is cold outside, are easily identified by the water vapour as a mist, referred to as plume, discharging from the boiler terminal.

How do these boilers extract all this extra heat? Basically, the boiler has a larger and more tightly grouped heat exchanger or, in some designs, such as the one illustrated in figure 3.15, has a second heat exchanger through which the flue products pass. The heat exchanger is the part that contains the central heating water over which the hot products of combustion pass.

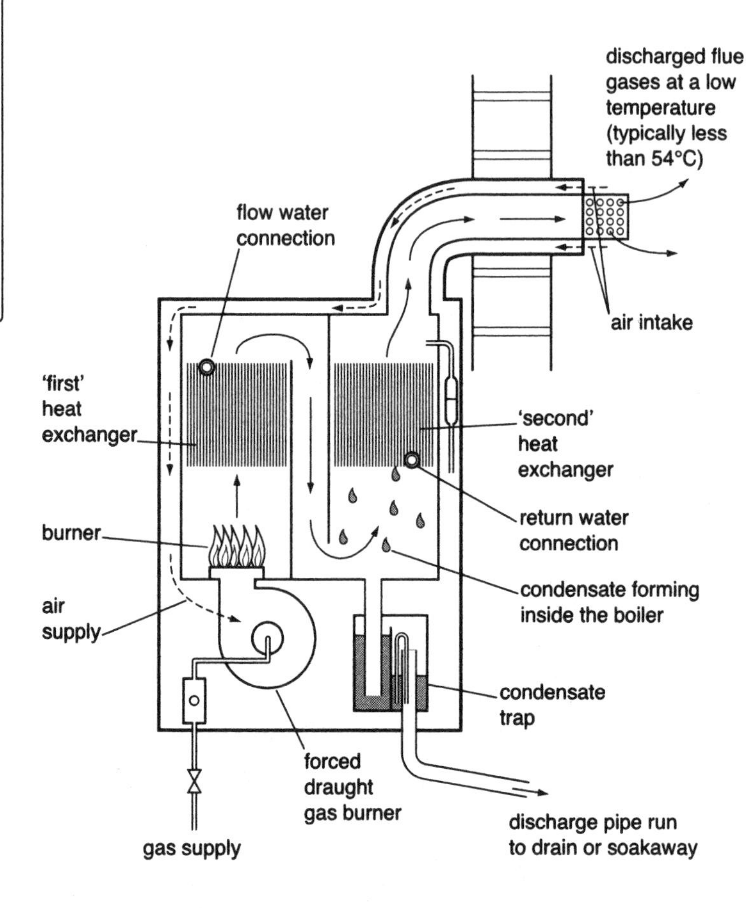

figure 3.15 internal view of a condensing boiler

High-efficiency boilers (HE boiler)

Of the boiler designs identified, those that work to the highest standard are of the forced draught room-sealed type. A modern boiler has electronic ignition and a highly efficient heat exchanger, making it far superior to an old cast iron boiler installed 30 years ago operating on gas with a permanent pilot flame that acts as the ignition source for the boiler. This old boiler might be operating at about only 50–60 per cent efficiency, whereas the modern boiler could be operating at efficiencies of over 90 per cent.

When talking of efficiencies, one is effectively talking of the running cost. For example, for every £100 spent on fuel, if your boiler is only 55 per cent efficient you will be getting only £55 worth of fuel and £45 would simply be going up the chimney. Likewise, where your boiler is 90 per cent efficient you will be getting £90 worth of fuel for every £100 spent.

It is because older boiler designs waste fuel in this way that current regulations no longer permit them be installed. If you need a new boiler the chances are, with a few exceptions, that the heating installer will be bound by law to install a boiler currently meeting efficiency of 86 per cent or above.

Heating controls

In your home you may or may not have all of the controls listed here; in fact, you may have no more than a switch to turn the power on to the boiler and pump. However, the design of a modern central heating system will use a whole range of controls in order to provide an efficient system. A requirement of the current Building Regulations for all new and replacement systems using gas or oil as the fuel source is to have a minimum of the following controls:

- full programmer or independent time switching for heating and hot water
- room thermostat, providing boiler interlock
- cylinder thermostat (where applicable), providing boiler interlock
- TRVs on all radiators, except in rooms containing a room thermostat
- automatic bypass valve (if necessary).

These controls all serve the function of reducing the amount of fuel required to heat the water and thereby increasing the efficiency of the boiler. In other words, they save fuel. Should you need to undertake any major renewal work in your home, such as to replace the boiler, your system controls will need to be upgraded as necessary and include all of the controls listed above.

With the exception of the TRVs previously identified, what do each of the remaining controls do?

Full programmer

This is in effect a fancy clock. It allows the water to come on at specific times set by the occupant of the building. Modern installations require the use of what is referred to as a full programmer. This basically means that the heating circuit/s and domestic hot water circuit can be controlled independently, allowing separate time settings for the heating and hot water. Earlier designs of programmers did not have this independence, for example:

- mini-programmers allowed heating and hot water to be on together, or hot water only (but not heating only)
- standard programmers allowed heating and hot water to be on on their own, but used the same time settings.

These older time controllers will need to be replaced if the boiler is replaced, thereby complying with the current Building Regulations.

Room thermostat

A room thermostat is a device that heats up as the temperature of the room increases. When the temperature set by the occupant is reached, an electrical contact is broken inside the thermostat to switch off the electrical supply to the pump or motorized valve found on the pipe serving the heating circuit. With no electrical supply the water ceases to be pumped to this circuit.

The room thermostat is normally positioned in a living room/lounge at a typical height of about 1.5 metres, but not in a position where it will be affected by draughts or by heat from the sun shining through a window. The thermostat should not be located in a room where there is an additional heater, such as an electric or gas fire – the hall might be a good alternative.

It is essential that the room selected for the thermostat does not have a TRV fitted to the radiator within the room because if the TVR closes, the room thermostat will fail reach its operating temperature. The idea of incorporating the room thermostat is to close off the heating circuit when the desired temperature has been reached in the living room. If the room thermostat is off, provided that the cylinder thermostat is not activated, the boiler and pump will be turned off, thereby saving fuel.

Some older systems may not have a room thermostat and just have TRVs fitted to the radiators to control the flow. These systems would need to be upgraded should the boiler be replaced at some time in the future.

Cylinder thermostat

The cylinder thermostat is a device fixed to the side of the hot water cylinder, about one-third up from the base. It is set by the installer to operate when the top of the hot water cylinder has reached a temperature of around 60°C. As with the room thermostat, when the desired temperature is achieved the electrical contact is broken inside the unit, which switches off the electrical supply to the motorized valve on the pipe circuit to the cylinder heat exchanger coil. Older systems may not have a cylinder thermostat. This is a situation that would need to be rectified should the boiler or cylinder be replaced, bringing the system in line with the mandatory regulations now in force.

Boiler interlock

Boiler interlock is when the boiler interlocks with a thermostat system so that it does not have a constant power supply. It will only ignite if heat is required by the domestic hot water or central heating system, which will be regulated by a room or cylinder thermostat.

Older systems did not always have a room or cylinder thermostat. For example, central heating systems were often designed only with TRVs fitted to the radiators, and gravity circulation of hot water to the cylinder from the boiler was allowed to continue until the boiler thermostat was satisfied.

Sometimes, to prevent the domestic hot water becoming too hot, a mechanical thermostat was installed in the return pipe to close off the flow of water in the circuit, and the boiler thermostat was the only control to switch the boiler on or off.

Invariably it continued to heat up and cool down all night and day as the boiler slowly lost its heat to the surrounding atmosphere. This is referred to as short cycling and is clearly a drastic waste of heat and fuel which boiler interlock prevents.

Systems without boiler interlock need to be upgraded when major work is undertaken on the system, such as when replacing the boiler. Where you only intend to replace the hot water cylinder you must include a cylinder thermostat to operate a motorized valve to close off the circuit and switch off the boiler, but you do not need to upgrade the central heating controls. However, if you replace the boiler both the cylinder thermostat and room thermostat must be provided, thereby providing total boiler interlock.

Automatic bypass valve

This device is a valve, fitted in a pipeline, that opens automatically to allow water to pass. There are several reasons why they are sometimes incorporated in the pipe circuit, such as because the boiler has a pump overrun facility. This facility is needed in systems where the pump must continue running after the boiler has switched off in order to allow the heat within the boiler to dissipate and cool down sufficiently, thereby preventing heat damage to the boiler itself.

If the motorized valves of the central heating circuit and domestic hot water circuit are open they will allow the water to flow, but where these are closed, due to the temperatures of their circuits being satisfied, there will be nowhere for the water to flow. As a result, pressure will build up within the flow pipe from the boiler and this will press against the spring-assisted valve of the automatic bypass to force the valve to open. Some boilers come with a pre-installed automatic bypass.

Prior to the automatic bypass, a slightly opened manually set lockshield valve was installed by the plumber, but this method is no longer permitted to serve this function as it can reduce the efficiency of the system.

Motorized valves

Older central heating systems will not have these controls as, prior to the 1980s, systems generally were installed as shown in figure 3.3. These older systems either had TRVs fitted to all but one radiator on the system to control the room temperature, or a room thermostat was used to control the heating requirements, which switched off the pump when the temperature within the room where the thermostat was located reached the required level. The temperature of the domestic hot water was generally only regulated by the boiler thermostat. These earlier systems, of which many thousands are still in existence, are far less efficient than the modern well-designed systems that use a motorized valve to close off the water supply to a particular circuit.

Closing off the motorized valve by way of the electrical power supply, from the room and cylinder thermostat as appropriate, provides a situation where the boiler is prevented from firing unnecessarily. The boiler of the modern system cannot fire unless either the room or cylinder thermostat is calling for heat, as it is these controls that send the power supply to feed a motorized valve.

The motorized valve itself consists of a small motor positioned on top of a housing inside which a ball-shaped valve is moved by the motor, opening or closing the route through which the central heating or domestic hot water can pass. There are two basic designs of motorized valve:

- two-port (zone valve)
- three-port (either mid-position or diverter valve).

When power is supplied to the motor in a two-port valve, it turns and causes the pipeline to open. As the valve opens it makes the switch contact inside the unit to allow electricity to flow to the boiler and pump. Should the power to this control be switched off the valve closes, assisted by a spring, and in so doing breaks the electrical contact to the boiler and pump.

There are two basic types of three-port valve: the diverter valve and the mid-position valve. The older designed diverter valve allowed the water to flow either from the central inlet port to the outlet pipe feeding the domestic hot water circuit or to the central heating circuit. In effect, it opened one route but closed the other, i.e. diverted the water flow, hence its name. This system was wired up to give priority to the domestic hot water in the cylinder, so that while this was being supplied with heat from the boiler the central heating system had to be off. This was affected by an internal ball which pivoted on a fulcrum between the two outlet ports.

The second type of three-port valve, referred to as a mid-position valve, allowed the internal ball valve to stop in the mid position as it was swinging across to close off one of the outlets, thus allowing water to flow to both the heating and domestic hot water at the same time, should both the room and cylinder thermostats be calling for heat. This mid-position valve therefore had the advantage over its predecessor diverter valve. It must be understood, however, that while the valve is in the mid position the amount of water flow that can be expected through the valve is restricted, so they are only suitable for systems that are not too large.

As for the operation of these valves, this is a lot more complicated to explain. The main thing to remember is that if no power is allowed to go to the valve the power cannot continue to travel to the boiler and pump.

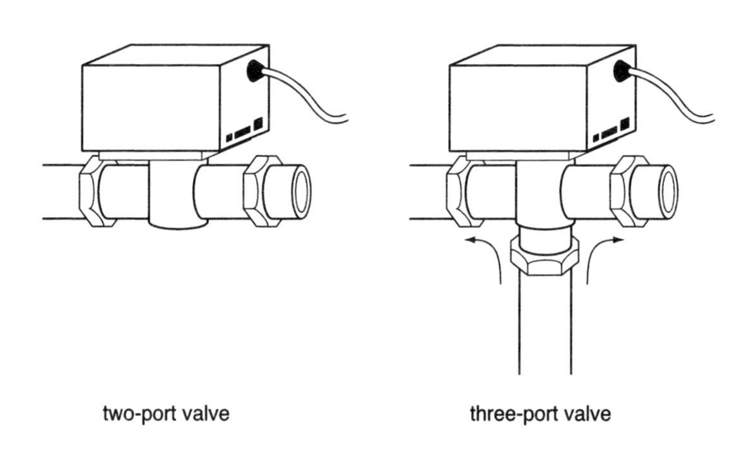

figure 3.16 motorized valves

Protecting heating systems

Frost protection in heating systems

Sometimes, where the pipework or boiler is found within an unheated part of the building, such as a garage or possibly the roof space, or where a separate outbuilding has been used for the boiler, it will be necessary to provide some form of protection against frost damage, including:

- filling the system with special central heating antifreeze
- using a special frost thermostat and pipe thermostat, positioned at the predicted coldest points in order to bring on the boiler with the intention of heating the water within, thereby maintaining it at a temperature just above the freezing point of water (0°C).

You will note that two thermostats have been listed, i.e. the frost and pipe thermostats and that they are used in conjunction with each other.

- The frost thermostat is designed to make its electrical contact due to the drop in outside air temperature.
- The pipe thermostat allows electricity to flow through its contact only where the water temperature inside the pipe, on which it is positioned, drops to around 5°C.

Thus, when it is very cold outside the frost thermostat makes its electrical contacts, which allows the electricity to flow to the pipe thermostat. If the water inside the pipe is sufficiently warm the electricity will not flow beyond this point, but if the water inside the system is dangerously cold it will allow the electricity to pass to the boiler and pump. Once the pipe thermostat is satisfied, with sufficient heat detected within the pipe, it breaks the electrical circuit.

Corrosion inhibitors

Corrosion inhibitors can be added to the central heating system in order to assist in prolonging its estimated lifespan. Several trade brands can be purchased from any plumber's merchant. The corrosion inhibitor serves several functions, including:

- lining the pipework in order to minimize the problems of corrosion
- lubricating the pump
- reducing the formation of bacteria within the system.

The only problem is the fact that, to have any real effect, it must be added to the system within a short time of installing the system.

You can also purchase cleaning solutions that assist in cleaning the pipework internally. These can be administered to some older systems prior to adding the corrosion inhibitor. You will need to obtain the manufacturer's data sheets to assist you further should you wish to consider treating an existing system – over zealous treatment could find leaks in your system that did not seem to be there prior to treatment. This is not because these destroy the pipe materials but because they destroy the sludge that has formed within, and it may be this sludge that is preventing a particular leak.

04 emergencies and contingency work 1

In this chapter you will learn:

- how to turn off the water supply
- how to drain the water from the system
- how to cure problems with leaking taps
- how to sort out problems with your toilet.

This chapter aims to look at some of the tasks that you may need to carry out in the event of something going wrong with your plumbing system. If you require expert advice or the services of a professional, see **appendix 3: taking it further** for trade and professional bodies.

Turning off the water supply

It is *essential* to make sure that you know where to turn off the water supply in an emergency. Make sure the valve works!

To turn off the internal cold water supply stopcock:

1 Find the valve (see Chapter 01, figure 1.2).
2 Turn the handle in a clockwise direction.

If it operates freely, then:

3 Continue turning clockwise.
4 Open the kitchen sink tap to check that the water has stopped flowing.

It is advisable to check that this valve works before an emergency arises. Simply try closing the valve as you would any other tap in the home. This will involve turning the operating handle or head in a clockwise direction. It is a good sign if the valve operates freely. Continue turning clockwise, counting the number of turns, until you feel the valve is fully closed. Then check that it has operated correctly by going to the cold sink tap in the kitchen and turning it on to see if the water stops flowing.

The kitchen tap is chosen because it is sure to be on mains supply, unlike other downstairs cold taps that may be fed via a cold water storage cistern in the roof space. When you try this tap, be prepared for the water to continue flowing for a short period before stopping completely because water may be draining out from the cold supply pipe within the house. Hopefully the supply will stop completely. If it does, there shouldn't be any problems. However, if the sink tap continues to drip you may need to apply a little more turning force to the supply stopcock to force the washer inside this valve tighter onto the seating.

To re-establish the water supply you simply open the stopcock, turning it anticlockwise the same number of turns as you counted when closing the valve. Finally, check to see that the water is flowing freely from the sink tap outlet. It might spurt out at first, due to the air pressure build-up caused by the air in the pipe compressing when the water is turned on; this is quite normal.

If all has gone well you have completed your very first plumbing job! Simple really, wasn't it?

Why did we count the number of turns when turning off the supply? This will be explained in more detail later (see page 134), but basically it is to ensure that you do not create any noise problems in your system. For example, if the supply stopcock was originally only open two turns and you then closed it and re-opened it by, say, four turns, you would allow a potentially greater volume of water to flow through the valve. This might cause shock waves to form within the system, due to such a large volume of water stopping when a tap in the system closes. These shock waves can create banging noises within the pipework.

One final point to note regarding the stopcock is that it is never a good idea to fully open the valve so that the head will not turn anticlockwise any more, as this means that the valve spindle is wedged up to its highest position. This might lead to the valve seizing up so, if ever you do require the maximum possible flow through the valve, open the valve fully and then turn it back half a turn.

To recap:

- turn clockwise to close the valve (and stop the water flow)
- turn anticlockwise to open the valve (and restart the water flow)
- count the number of turns when opening or closing the valve
- keep the valve labelled up for easy identification
- ensure that the valve is operated occasionally to ensure it continues to work freely.

Problems with turning off the supply

Unfortunately, it is not always as straightforward to turn off your supply as just described. Often, due to insufficient maintenance and lack of use, typical problems such as the following are encountered:

- No movement of the tap head due to it being seized up.
- Water leakage past the spindle after the valve has been operated.

- It continues to flow when you have turned off the valve.
- There is no water flow when the stopcock is re-opened.

These points are discussed in greater detail below.

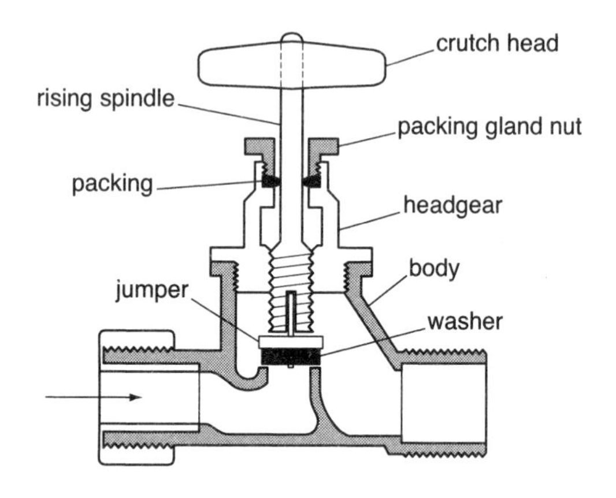

figure 4.1 section through the supply stopcock

Main stopcock seized up

Since the valve is, as a rule, not operated year in and year out it simply seizes up due to lack of use. There is only so much torque that you could apply to the head before some damage would result, so what can be done to help?

You can try loosening off the packing gland. Referring to figure 4.1, take a small spanner and slightly undo the packing gland nut by turning it anticlockwise. This releases some of the pressure on the packing inside the spindle. The packing is designed to prevent water seeping past the spindle itself when it is turned. This may be all that is required, but it may be necessary to undo this nut substantially before any movement of the head is possible. It is not advisable to completely remove this packing gland nut because until it is possible to close off the supply, water could leak from this point, and you may want to retighten the gland nut to stop the leak.

If the valve still will not budge you will need to consider turning off the outside stopcock or contacting the water supply authority to request that they turn off your supply for repair. With the water to this seized up valve turned off, the water will need to be drained from the system so that the valve can either be stripped down to free up the component part, or replaced completely.

Water leakage past the spindle of the stopcock

After operating a stopcock that has not been used for some time, occasionally you will find that when you turn back on the supply, water seeps past the packing gland nut. You may have had to loosen off this nut to close the valve in the first place, causing this leak. What can be done? Simply try to tighten the gland nut by turning it clockwise.

Tightening this nut applies pressure onto the packing, squeezing it out to form a tighter seal. This can unfortunately have the effect of making the valve very stiff to operate, and sometimes just tightening this nut is not sufficient to cure the problem, in which case you may need to repack the gland.

To repack the gland:

1. Fully close off the stopcock that is to be worked on.
2. Completely undo the packing gland nut by turning anticlockwise until it comes away from the housing and can be slipped further up along the spindle. Very little water should come out because you have turned the valve off and any water will be the result of it draining down from the system. You may need to drain this water via the drain-off cock, located above the stopcock.
3. With the packing gland nut removed, wrap a few strands of PTFE tape (an abbreviation for Polytetrafluoroethylene, a common plastic jointing material available from all plumbing supply outlets) around the spindle and push it into the void into which the packing gland nut screws, poking it down with a small screwdriver (see figure 4.11).
4. Now replace the packing gland nut, tightening it just sufficiently to squeeze the new packing material within the gland.
5. Re-open the valve and tighten the packing gland nut until the water seepage past the spindle stops.

Stopcock ineffective when closed

If you have turned off the main supply stopcock and water continues to flow from the kitchen sink tap it is likely that the washer has perished and no longer functions. The first thing to do is to double check that the valve is fully closed and not just stiff. Having done that, you should consider turning off the outside stopcock or contacting the water supply authority requesting that they turn off your supply for a repair to be initiated.

With the water turned off and the water drained from the system it will be possible to strip the valve down to re-washer it. This can be done as follows:

1 Use a spanner to undo the headgear, as seen in figure 4.1, by turning anticlockwise. This removes the top half of the valve from the body attached to the pipe.
2 With this removed, you will be able to remove any remains of the old washer and replace it with a new ½" tap washer, obtainable from any plumbing suppliers.
3 When you replace the headgear, check that the fibre washer used where the head meets the body is in good condition; if it is not, water may escape from the joint where the two surfaces meet. There is usually no problem with this fibre washer, but occasionally they perish. Usually a few turns of PTFE tape between the mating surfaces, forming a new washer, is all that is needed to form a tight seal.
4 Turn back on the water supply and test this valve for correct operation, ensuring it does not leak past the spindle or from the body of the tap where you removed the headgear.

No water flow when the stopcock is re-opened

This is another problem that can occur when you turn off the water supply. It is caused by the washer becoming detached from the jumper (see figure 4.1) and remaining stuck down onto the valve seating. You can try giving the side of the tap a knock in the hope of dislodging it, but in reality the supply will need to be turned off and the valve stripped down to re-washer the valve.

Turning off an external underground supply stopcock

As you may have gathered from the solutions identified above, it is generally inadvisable to access the outside stopcock unless you are fully prepared to dig up the ground around the stopcock to expose it. However, if there is an emergency and you need to turn off the supply at all costs you may need to do this. The valve will be at a minimum depth of 750 mm below ground and may be even deeper, so you will need to use a stopcock key to access the valve. This is designed to pass down the large pipe duct leading to the stopcock and slip over the top of the valve head to initiate the turn.

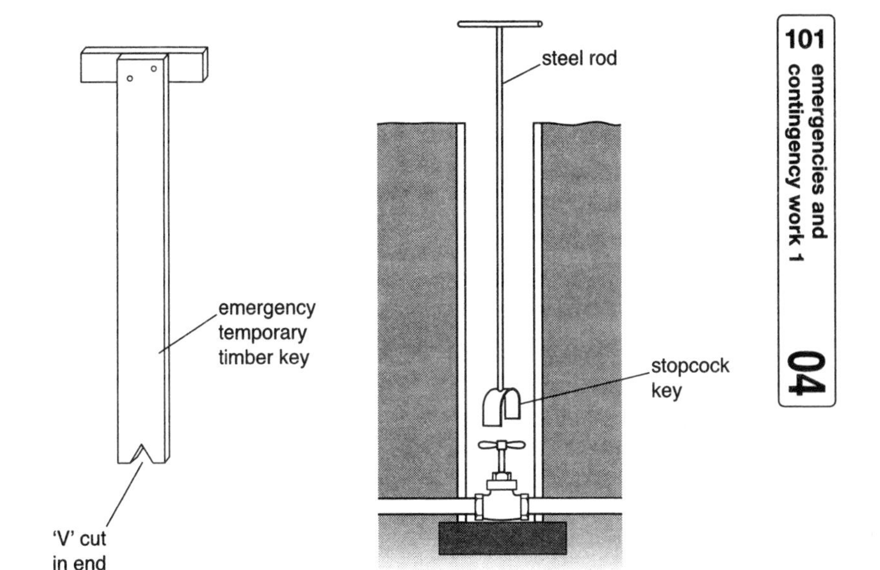

figure 4.2 using a stopcock key

Turning off the cold water from a storage cistern (low-pressure pipework)

Where the water feeding an appliance such as a sink, bath or toilet cistern is supplied directly from the mains inlet, turning off the incoming supply stopcock will stop the flow of water. However, if the water continues to flow you will know that it is being supplied via the cold storage cistern.

If you simply turned off your incoming supply stopcock and waited long enough the water would eventually stop flowing, as the cold storage cistern would eventually empty. However, looking at figure 1.4 you will see that there is a stop valve located in the pipeline as it exits the storage cistern, so turn off this valve by turning the head clockwise.

This valve may be located in the loft close to the storage cistern or, if you follow the route of this pipe where it passes through the ceiling to drop down to the room below, it may be found in a cupboard for convenience. Some systems, such as those used in blocks of flats, do not have a loft and therefore have the cistern located high in a cupboard somewhere around the property. This valve will not be of the same design as the mains supply stopcock. Traditionally gatevalves have been used here, although due to their unreliability they invariably have been replaced in recent years with lever-operated quarter-turn valves.

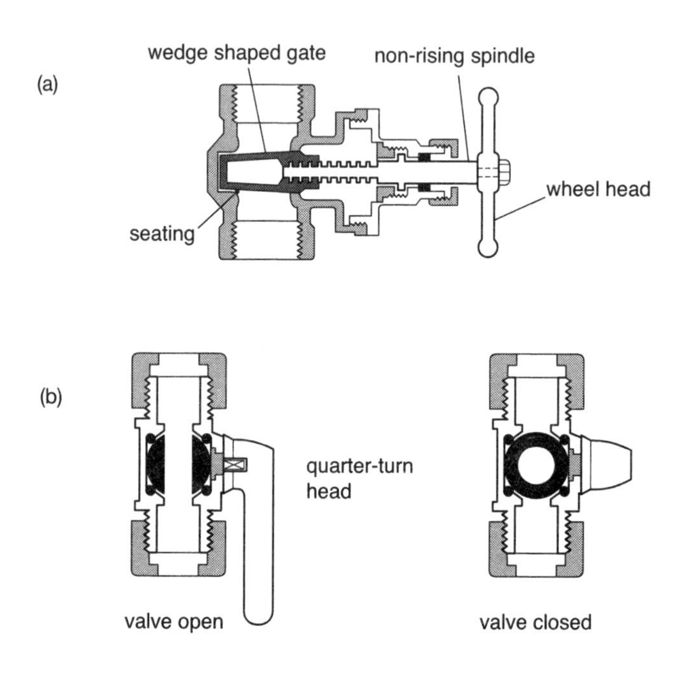

figure 4.3 gatevalve (a) and quarter-turn lever-operated valve (b)

Where a gatevalve is found, it may only halt the main flow of water and still let a little flow by. As I said, they are not always very effective. Also, sometimes they fail to operate and even when they do they sometimes fail to re-open. If a quarter-turn valve is found, simply turn it one-quarter of a turn until the handle is perpendicular to the pipe. These do not usually cause a problem and therefore I would always recommend this type of valve is installed should you consider a new or replacement valve. Where a new valve is sought, the type selected must maintain a full bore when looking through the valve in the open position and not be of a design that offers a reduced bore, or you will notice the lack of water flow when it is installed.

Where you cannot locate the valve from the storage cistern or it does not effectively work, it is possible to block the outlet pipe from the cistern with a cork. Alternatively, turn off the supply feeding the cistern and drain out the water via the taps fed from it. The cistern can be turned off at its inlet stopcock or screwdriver-operated quarter-turn valve. Where you cannot find a valve you can lift the arm of the float-operated valve, which will stop the water flowing into the cistern. You can tie this up using a piece of string and a batten resting across the top of the cistern.

If the storage cistern feeds the hot water supply as well as the cold water, draining out the water from the cistern will also stop the flow of water from the hot taps.

Turning off the hot water supply

Following the same principles as for turning off the cold water supply, you need to go upstream of the hot water heat source to locate an isolation valve on the pipework.

In the case of a combination boiler this will be a quarter-turn valve found just beneath the boiler itself. This valve may have an operating handle or you may need to use a spanner or screwdriver to turn off the valve, giving just one-quarter of a turn. For other instantaneous systems there may be a valve incorporated with the appliance or you may have to source a valve on the pipework to the appliance.

In stored hot water cylinder systems you will find the isolation valve on the pipe supplying the cylinder (see figures 2.6 and 2.8). This stop valve may be in the same cupboard as the cylinder or you may need to go into the roof space, where a vented system is supplied by a cistern. With the water supply to the hot water cylinder isolated it is also advisable to turn off the power supply to the heat source.

With the supply to the cylinder closed off, water can no longer enter the cylinder and therefore when a tap is opened on the hot distribution pipe from this vessel it runs only for a minute or so, just sufficiently long to drain the water from the leg of pipework from the cylinder to the tap. The cylinder itself remains full, even though no water flows from the tap. If you look at any of the examples of stored hot water supply discussed above, or figure 4.4, you will notice that the hot water is drawn off from the top of the cylinder and when the supply to the cylinder is closed, water is trapped in the cylinder within a big U-shaped leg of pipework.

To remove this water, for example to replace the cylinder, the drain-off cock at the base of the cylinder will need to be opened, draining the water through a hosepipe to an outside drain. Figure 4.4 also shows the water that is trapped in the piece of pipe prior to the outlet of the tap, which again has been trapped within a low section of pipework. You need to be aware of this when working on any drained-down pipework as when, for example, cutting into the pipe, water will flow from it until it is all drained off. This can be a little disconcerting for the novice plumber who has turned off the water supply, checked that nothing comes out of the taps and so proceeds to cut the pipe!

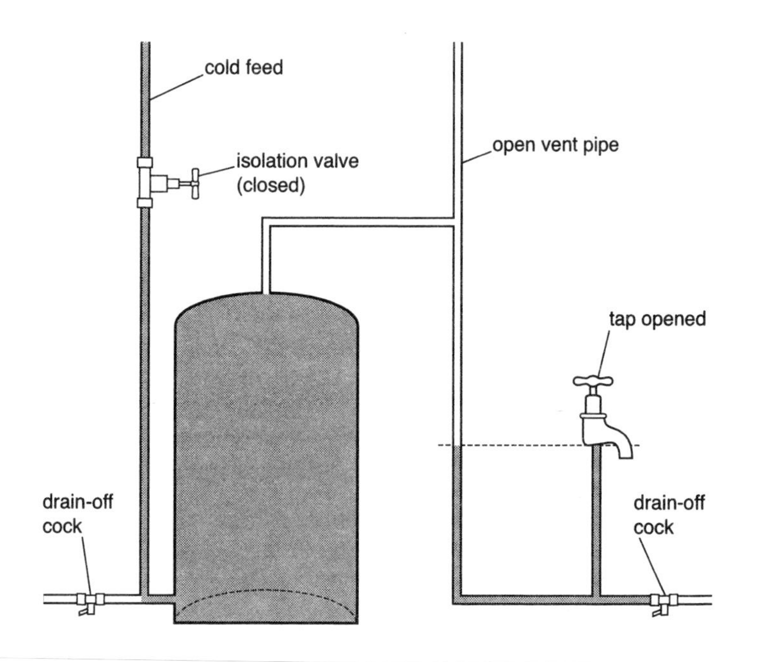

When the isolation valve is closed, water will stop flowing from the tap, but water is still lying within all the areas shaded and these will require draining via the drain-off cocks if you wish to cut into these parts of the system.

figure 4.4 water trapped within the pipework

Draining water from the various systems

Draining down the hot or cold water supply

Draining down either the hot or cold water systems will only be necessary when you wish to undertake a major repair or alteration work, or if you plan to go away for an extended period in winter leaving an unheated building. When going on holiday, especially during the summer, it is normally just enough to turn off the incoming water supply as a precaution.

In order to fully drain down your pipework you will need to:

1 Locate the supply isolation valve and close it. If you plan to drain down everything this will clearly require the mains supply stopcock to be turned off. However, if it is just, say, the low-pressure cold or hot water that needs draining down then you will need to close the valve from the storage cistern. If the cold water within the storage cistern itself needs to be drained then the inlet valve to the cistern must be closed.
2 Open the tap fed from the isolation valve that you have just closed. Water will flow until it has drained from the pipe. At this point it must be understood that although no more water is feeding the system there will still be water lying in pockets of pipework, and in the case of the hot water system the pocket includes a cylinder full of water (see figure 4.4).
3 Connect a hosepipe to the drain-off cock (see figure 4.5) and run it to a suitable discharge point. The square-headed drain-off valve can now be turned anticlockwise to open it, with the aid of a spanner, allowing the water to flow. However, do not completely remove the spindle of this valve as it unwinds, otherwise the water will discharge all over the floor.
4 Open several outlets to help with draining the water in both 2 and 3 above, and flush the toilet if it is fed by the section being drained – it allows air into the pipework which will help to remove the water.

Remember: Where a hot water system is drained down the heat source must also be isolated to prevent damage.

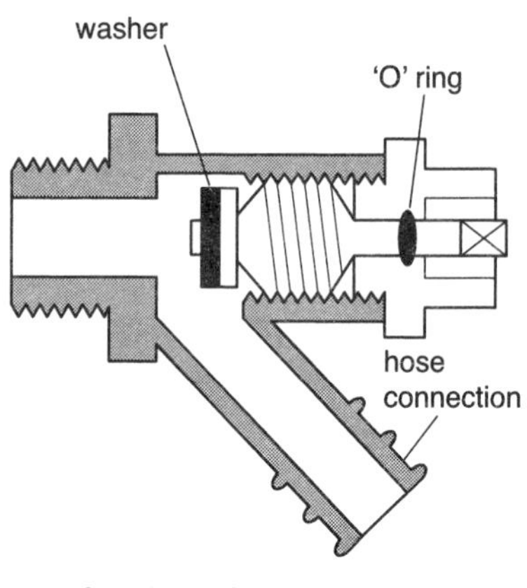

figure 4.5 a drain-off cock

Draining down the central heating system

Should you ever need to drain down the water from the heating system, this is how to do it:

1 First, turn off the power supply to the central heating system to ensure that the boiler cannot fire up.
2 If it is a vented system you must now turn off the isolation valve supplying the system. This will not apply for a sealed heating system because there will be no permanent water connection to the system. Even if the temporary hose that was used to fill the system had not been removed by the installer, the isolation valve would still be in the closed position. The isolation valve for the vented system can be found on the inlet supply pipe to the feed and expansion cistern, located in the roof space. If there is no valve you will need to get a piece of wood, position it across the top of the cistern and use a piece of string to tie up the lever arm to the float-operated valve, thereby preventing it from opening as the water in the system drains out.
3 You are now ready to start draining down. Locate a drain-off cock (see figure 4.5) somewhere on the system at a low point.

There is usually one situated near the boiler itself. Connect the hose and open the valve. The water flow may be a little slow at first as air needs to get into the system in order to let the water out.

4 Now go to a radiator, high in the system, such as on the first-floor level, and open the air-release valve with a radiator key – this will assist the process of draining down by letting air into the system. You will hear the air rushing into the system as the water empties out.

5 Slowly open more radiator air vents, doing the higher level radiators first, until all of them are open. It is essential to keep an eye on the water draining from the system as the drain-off cock is renowned for letting water escape through its thread, so you may need a flat tray suitably positioned to catch the water.

6 When all of the water has been removed it is wise to close the radiator air vents to ensure that the system is ready to refill and to make sure the valves, if fully removed, are not lost.

Sometimes the drain-off cock fails to open simply because the washer is stuck onto the seating. The best thing to do is to try and locate another valve. However, if you need to open the one that is stuck then you could try tapping the side of the valve, or you may need to totally remove the screw-in spindle and poke a small screwdriver into the valve to dislodge the washer. If you do this you must be totally prepared for the water that will gush from the fitting once the washer is dislodged, and be prepared to re-insert the valve head. Sometimes you can attack this washer through the hose connection hole. Take note: the water drained from a heating system can be just like black ink and stains as much, so do not take this course of action unless you definitely know what you are doing. Finding another drain-off cock may save a lot of hassle.

Drain-off cock failing to close off the water

When draining down any system, once the work is completed it is always worth removing the spindle completely to look at the state of the small washer. Often these are perished and fail to keep the water within the system when they are reinstated. So, to avoid having to re-drain the system later to replace this washer, inspect it now and replace the washer if necessary. See figure 4.5.

A dripping tap

A dripping tap is one of the most common problems encountered in the home. It is the result of one of the following:

- faulty or worn out washer
- dirt or grit lodged across the seating
- damaged seating
- damaged ceramic discs (quarter-turn taps).

Replacement of the tap washer

Replacing a tap washer is a relatively simple process. The first thing to do is to turn off the water supply feeding the tap, then open the tap as far as possible to make sure no water flows from it before proceeding. If you look at figure 4.6 you will see that there are a few differences in the general designs of taps. Basically, taps will have a rising or a non-rising spindle.

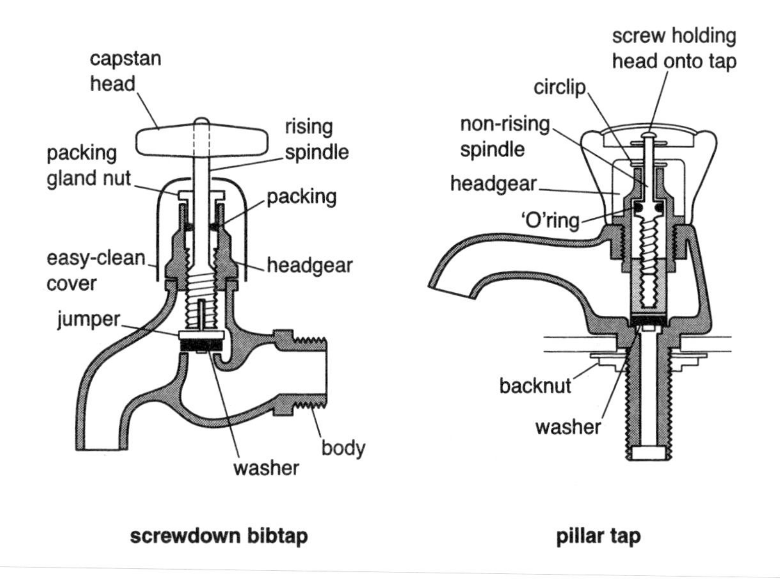

figure 4.6 types of tap design

Taps with a rising spindle usually have chromium plated easy-clean covers, which will need to be removed in order to get a thin spanner onto the nut of the headgear underneath. To remove this cover, grip it firmly and turn it anticlockwise to unwind it. You may need a spanner for this process, but take care not to damage the chromium finish. You will not be able to completely remove this cover because of the capstan head, but you should be able to lift it sufficiently to get your spanner in to grip the headgear itself, which is also turned anticlockwise to remove it from the tap body. When doing this, the body of the tap needs to be held very firmly to prevent it from turning within the appliance.

Taps with a non-rising spindle incorporate an aesthetically designed easy-clean shrouded cover, which must also be removed in order to gain access to the nut of the headgear. This may simply pull off, but it is usually held on with a small fixing screw holding it in position. This will be either somewhere around the perimeter of the operating handle or, more likely, beneath the tap indicator cover found on top of the tap as seen in figure 4.6. To remove this cover, usually a small screwdriver is needed to ease it up. With the cover removed you will see the screw that holds the top of the tap in place; this can be undone and the operating handle pulled from the tap. A spanner can now be used, as described above, to unwind the head of the tap from the body.

With the head finally removed from the body of the tap, you will see the washer on the base of the jumper. Look inside the body of the tap to inspect the surface of the seating and check that there are no obstructions preventing effective operation. With the washer located it can simply be removed and replaced with a new one. The washer may be pressed onto a small central stem or it may be held in position by a small nut. If the nut is difficult to undo it will need to be soaked in penetrating oil to free it because, if the nut were to snap off, the whole jumper would need to be replaced. Taps on baths use a ¾" tap washer, whereas all other appliances use a ½" tap washer. With the washer in place the procedure is reversed, reassembling the tap and testing to check that it now effectively closes off the supply.

So, to recap:

- turn off the water and ensure the water is off
- remove chrome shield from tap
- using a spanner, remove the headgear from the tap body
- remove old washer and fit on the replacement
- reassemble and test.

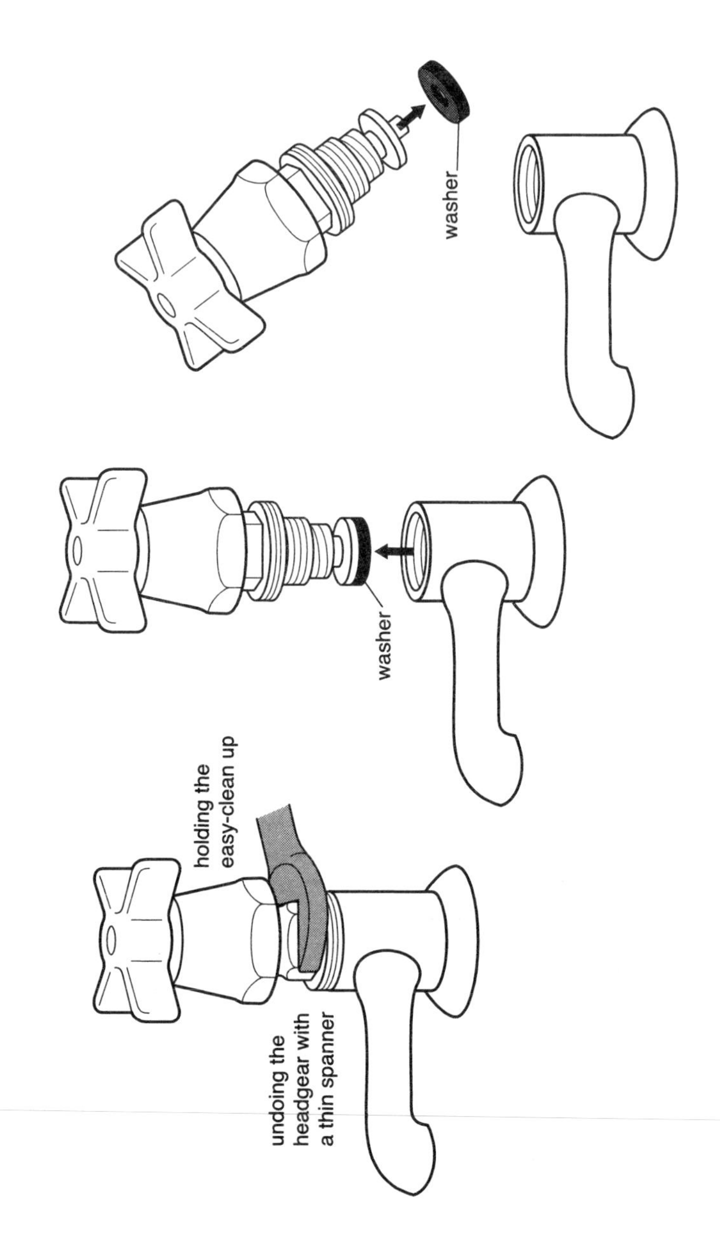

figure 4.7 re-washering a tap

Tap is still dripping!

Having put a new washer into a tap and discovered it still drips suggests a far greater problem. First, a second washer could be tried, ideally a softer one. However, it might be that the seating has become eroded. This can occur, particularly where the pressure is very high. The answer is to do the following:

1. Install a nylon substitute seat (sold with a matching washer). This is dropped over the old seat and as the tap is closed it is forced into position.
2. Re-cut or smooth off the original brass surface seating where the washer sits. In order to do this you first need to obtain a tap reseating tool. This should be obtainable from a reputable plumber's merchant for a few pounds.

Using a tap reseating tool

With the water turned off and the head removed, as above, the tap reseating tool is screwed into the body of the tap. The tool is adjusted and wound down until the cutting head reaches the seating. The handle of the cutting head is now turned a few times to cut off a thin layer of the brass seating. Once removed, look inside the tap to inspect the seating and, if it looks okay, reassemble the tap with a new washer.

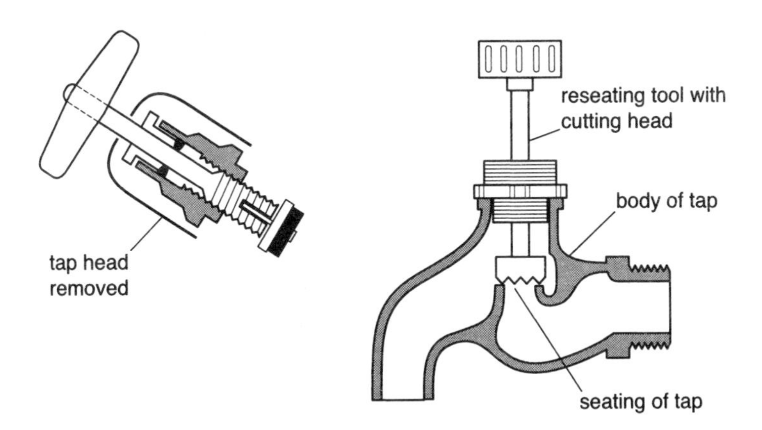

figure 4.8 reseating a tap

Re-washering Supa-taps

Although these taps are no longer made there are still vast numbers in existence. They were designed in such a way that it is possible to re-washer them without turning off the water supply, as follows:

1 First, hold the operating head of the tap firmly and unwind the retaining nut by turning clockwise with a spanner (it is a left-handed thread). The head will not drop off, but when the tap is turned, as if opening the tap, it will eventually drop into your hand. At this point the self-closing device should drop to form a kind of seal stopping or greatly reducing the water flow while the washer is replaced. If it does not drop don't panic, the water will only flow into the appliance and it is possible to poke a small screwdriver up into the outlet to dislodge this closing device, thereby allowing it to drop. Alternatively you can simply turn off the water supply to this tap.
2 With the tap body in your hands the washer initially looks as though is cannot be reached, but by pushing the water outlet point against a block of wood the washer and its anti-splash device will pop out.
3 Now separate the washer from the anti-splash device by prizing the two apart with a screwdriver. It should be noted that the Supa-tap washer is encased with its own jumper and therefore needs to be replaced as a complete unit.
4 With the new washer in place the sequence of events is reversed to reassemble the tap.

Taps with ceramic discs

Dripping taps that use ceramic discs often require the replacement of the discs themselves. It is always worth looking to see if there is any grit or blockage preventing full closure of the valve, but when cracked or damaged they will need to be replaced. They are available from a supplier but will sometimes require special ordering, so you may need the precise details of the manufacturer and product type.

The discs are supplied as a cartridge and the cartridge for the hot tap turns in the opposite direction from that used for the cold supply, so ensure you fit the right ceramic cartridge type.

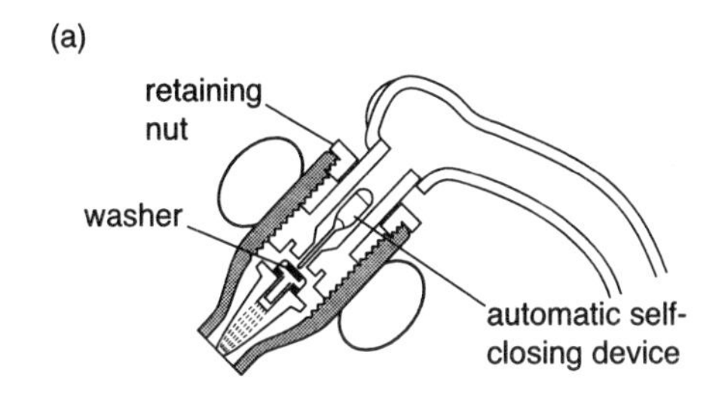

internal view of tap

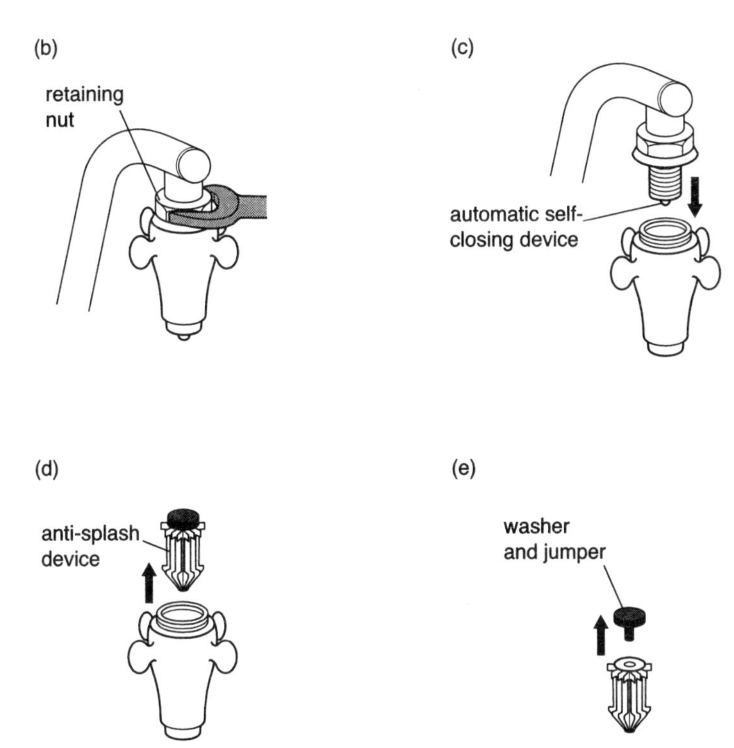

figure 4.9 re-washering a Supa-tap

In order to get to the ceramic discs you need to follow the procedure for stripping down the tap as detailed for the re-washering of taps – you will find the discs in place rather than a washer. At the same time as you replace the discs, replace any rubber sealing washer supplied.

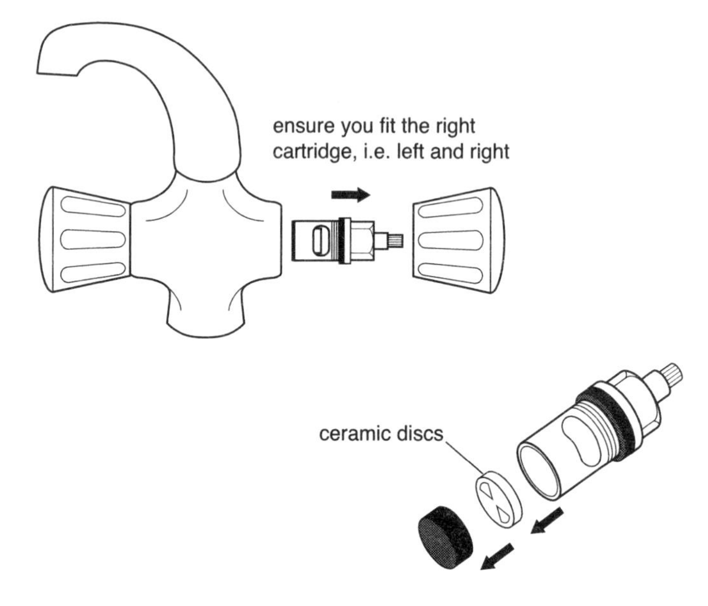
ensure you fit the right
cartridge, i.e. left and right
ceramic discs

figure 4.10 cleaning and replacing ceramic discs

Water leaking from the body of a tap

If you open a tap and water seems to escape from a point somewhere around the spindle, it is probably the result of water leaking past the gland where the spindle turns. The water will only leak when the tap is running. To resolve this problem you need first to identify the design of the tap, i.e. has it a rising or non-rising spindle (see figure 4.6).

Water leaking from the body of a tap with a rising spindle

With this design of tap there will be no need to turn off the water supply; you will just need to turn off the tap fully when you work on it. The procedure is straightforward once the easy-clean cover has been removed, but this in itself can be a tricky task as you might have difficulty in removing the capstan head because it may not have been removed since the tap was first installed. Here is some guidance for this task.

1 First, look for a small screw holding the capstan head on. Look around the base of the capstan head, or under the plastic red (hot) or blue (cold) indication marker on top of the handle. Occasionally there is no screw.
2 Holding the body of the tap firmly, now try to pull the capstan head off; if it is held on very firmly a few gentle taps with a small wooden mallet aimed upwards might dislodge it. One trick that can be used is to open the tap fully, with the easy-clean cover also undone and raised as high as possible, and insert a block of wood tightly between the cover and the body of the tap. The tap is now closed and with luck the process will have jacked the capstan head off from the spindle. Alternatively, the use of some penetration fluid might do the trick. The amount of force required can sometimes be quite excessive and on rare occasions removal of the tap from the appliance (e.g. basin) might be required to ensure that no undue damage is done.

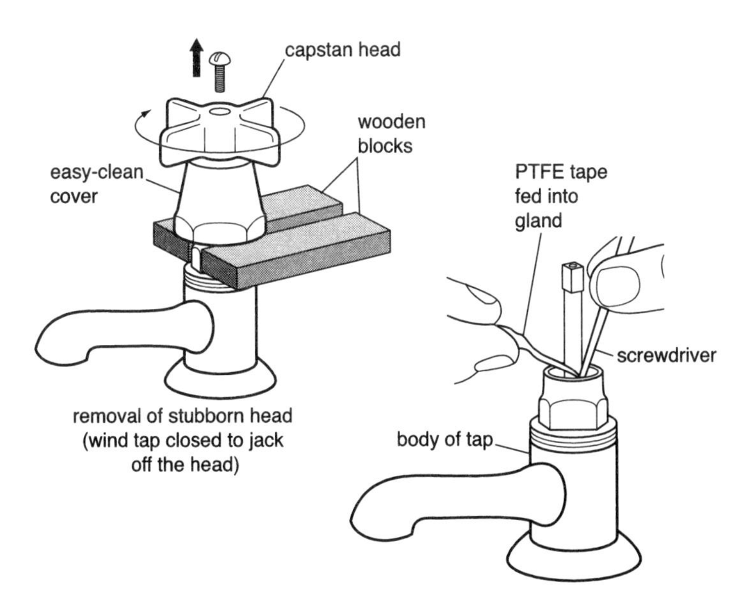

figure 4.11 repacking the gland

With the capstan head and easy-clean cover removed, you will be able to see the packing gland nut. You will see that when the tap is opened water will discharge from this point and will stop when the tap is closed.

Tightening up the gland nut a little may be all that is required. However, where this does not cure the problem:

1 Turn off the tap.
2 Unwind this nut and remove it from the spindle.
3 Pack a few strands of PTFE tape or some waxed string around the spindle and push it into the void into which the packing gland nut screws, poking it down with a small screwdriver. (See figure 4.11).
4 Replace the packing gland nut, tightening it just sufficiently to squeeze the new packing material within the gland.
5 Re-open the valve and, if necessary, tighten the packing gland a little more until the water stops seeping past the spindle.
6 Finally, reassemble the easy-clean cover and capstan head.

Water leaking from the body of a tap with a non-rising spindle

With this design of tap the packing gland has been replaced with a rubber 'O' ring (see figure 4.6). Once you have removed the tap operating head you will see that the water escapes past the spindle if the tap is turned to the 'on' position.

To cure this problem:

1 First, turn off the water to the tap.
2 Remove the easy-clean shrouded cover and remove the headgear from the tap body as described earlier (see page 109).
3 Now remove the circlip located at the top of the valve. Do this by placing a screwdriver between the open edges and twisting gently, thereby forcing it apart to slip from the spindle. Unfortunately this circlip will sometimes break, in which case it will need to be replaced (see figure 4.12).

4 With the headgear in your hands, push on top of the spindle, unwinding it and removing it from the brass housing, exposing the 'O' ring.
5 The old ring can now be flicked off, usually with the aid of a small screwdriver.
6 Replace the 'O' ring with a new one, applying silicone grease to provide a lubricant.
7 Now reassemble the tap and test it. If this repair does not resolve the problem it may be due to excessive wear of the spindle, in which case a replacement tap would be required.

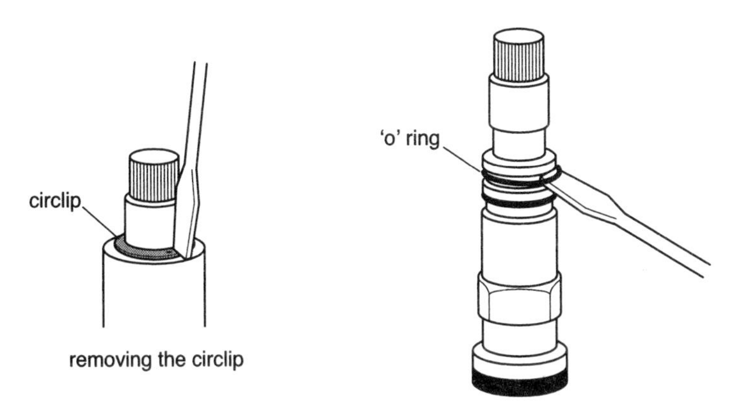

figure 4.12 replacing the 'o' ring in a tap with a non-rising spindle

Water leaking from the swivel outlet of a tap

This is the result of a worn out 'O' ring found at the base of the swivel spout. There is no need to turn off the water supply, just turn off the hot and cold taps. The first task is to remove the small retaining screw or locking nut located at the base of the spout (note that some designs do not have this securing device). You will then need to turn the swivel outlet to one side, in line with the tap heads, and pull it off from the body of the tap to expose the large 'O' ring. You can now replace this with a matching replacement, applying a little silicone lubricant as necessary. Where excessive wear has occurred the problem may still exist, in which case a replacement tap would be required.

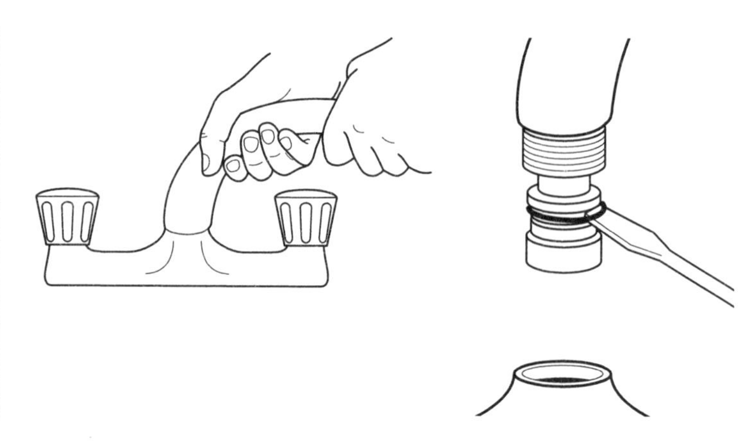

figure 4.13 replacing the 'O' ring in a tap with a swivel spout

Lack of water flow from a tap

The water that discharges from a tap should come out with sufficient force (i.e. the force you would expect from a normally working tap). Some taps are undoubtedly better than others, with a greater flow or pressure, but in general we know what to expect. Therefore, when faced with a tap that has a poor water flow rate you can surmise that something is wrong.

Before condemning the whole system of pipework, consider that it may be the tap itself that is faulty. Does it operate freely and open fully? Look at the pipework and see what else is served from the same section, and check out these taps or outlets too. Are they also suffering this lack of flow? If so, has the problem been getting slowly worse or is it a sudden drop in pressure or flow? Clearly if several taps are affected there is some form of blockage in the pipeline and this will generally be one of the following:

- turned off or closed down water supply
- a blockage due to debris in the storage cistern
- a plug of ice
- an air lock
- corrosion or limescale build-up.

The first thing to do in this situation is to look for the source of the water that is being blocked. If the problem occurs suddenly, affecting the cold water mains supply to the kitchen sink tap, possibly with the flow stopping completely, it may be worth phoning your water supplier as they may have turned off the water for some reason.

For your low-pressure pipework, such as that serving the hot and cold taps to the bathroom, check whether the storage tank in the loft is full of water. Is the lid in place, ensuring vermin have not got into the vessel, drowned and sunk to the bottom, blocking the outlet pipe?

The weather will be a sound indicator of whether a blockage is due to ice. This scenario is discussed later (see page 204).

The problems of air locks, corrosion and limescale, however, may not be so obvious and are discussed in more detail below.

Blockages due to an air lock

Where an air lock is suspected to be the cause of a lack of water flow to a tap, the air must be forced from its trapped location. An air lock is the result of poor plumbing design in a low-pressure (storage-cistern fed) system. When running pipework you should never run it uphill and then downhill as air will accumulate in the high pocket. With air trapped in the high pocket, if the water pressure is insufficient no water can pass. You should aim to design your system so that air within can always rise and escape through the tap outlet or via the cold feed or open vent pipe.

With a badly designed system with high points, once the air has been expelled from these high points it allows water to flow, so the problem generally only reoccurs when a system has been drained down. If you didn't install the plumbing system you won't know if it was installed correctly, therefore, as the unsuspecting individual who has drained down the pipework you have no idea that this will occur until you try to refill the system. So, let's say you have turned on the water supply after draining down for some reason and there is no water flow at the outlet. You simply do not know where the high point is causing the trapped air, so what do you do?

The first thing to try is to open the tap and, either using a hose connected to its outlet or positioning your mouth around the spout of the tap, try to give a good blow. What this sometimes does is blow a bit of water that was lying inside the pipe up, forcing itself past the trapped air which can then escape back out of the system, via the cold feed. A variation to this is to get a small length of hosepipe and pass it down into the storage cistern and into the cold water pipe feeding the offending section of pipework. With the hose in place and with the tap opened you can now blow with all your might in the hope of forcing the water past the trapped air lock.

Failing this, try a trick used by many plumbers. Get a small piece of hose pipe and join the cold high-pressure mains water tap outlet to the low-pressure tap outlet and use this water pressure to force the air from the high section. I must add that although this trick works, technically you are in violation of the Water Supply Regulations unless you have some means of backflow in place to ensure that contaminated water cannot flow back into the water authority mains.

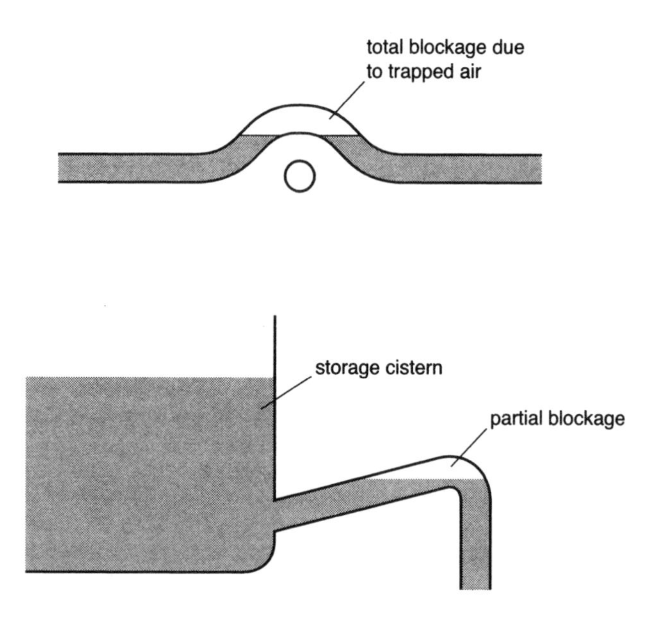

figure 4.14 common causes of air locks

Blockages due to corrosion or scale build-up

Limescale build-up has already been discussed, but recognizing it as a problem is not always easy. It is generally accepted that both corrosion and scale build-up problems occur over a long period of time, during which things slowly get worse.

The type of metals used will be an indicator of the likelihood of pipework suffering from corrosion. Galvanized iron pipes in particular can be a problem, and where these are encountered you should always suspect them of causing a reduced flow rate of water. For example, where galvanized iron has been used in conjunction with copper or brass, electrolytic corrosion occurs, creating encrustations within the pipe. The worst affected point would be where the two dissimilar metals join together. Fortunately, galvanized iron is no longer used for domestic pipework and therefore this problem will only occur in older dwellings. Electrolytic corrosion is discussed in Chapter 06.

A blockage due to limescale is not as easy to detect and will only be a problem in hot water systems. The point where the blockage may be worst is within the pipe exiting the top of the hot water storage cylinder. If you isolate the water supply to this cylinder and remove the pipe coming from the top dome of the cylinder to inspect it, you may find that limescale is blocking the pipe. This point is selected because it is where the hottest water will be found and therefore it is here that the limescale is most likely to form. (Figure 1.10 shows a fitting that has been blocked with limescale.)

Toilet will not flush

Siphon type

In order to flush the toilet the lever arm is operated; this lifts the large diaphragm washer inside the siphon tube (see Chapter 01). If the WC fails to flush, the first thing to do is simply to lift the lid from the cistern and check the operation of the linkage system used to lift up the diaphragm washer. Assuming this is okay, the fault will almost certainly be a split or worn out diaphragm washer. This can easily be replaced but, unfortunately, in the case of close-coupled WC suites, it requires you to remove the whole cistern from the wall in order to remove the siphon. There is a siphon design built as two parts, which allows you to pull the siphon apart to facilitate this repair, but unfortunately these are not commonplace. Toilet cisterns with flush pipes such as that shown in figure 1.20 do not need to be removed from the wall.

To replace the washer:

1 Turn off the water supply to the WC cistern – there may be a quarter-turn valve on the inlet supply pipe.
2 You now need to be bail out the water from the cistern, using a sponge if necessary to draw out every remaining drop of water, otherwise what remains will discharge onto the floor when the siphon is removed.
3 For cisterns with a flush pipe, unwind the large nut securing it to the siphon, turning it anticlockwise.
4 Next, unwind the big nut securing the siphon to the cistern.
5 The siphon can now be lifted from the body of the cistern. To complete this action you will need to unhook the linkage to the lever arm and sometimes, if the arm of the float-operated valve gets in the way, you may need to remove this as well.
6 With the siphon removed from the cistern you can now see beneath the base of it and will see the location of the old perished diaphragm washer.
7 Remove the hook attached to the top of the shaft that pulls the diaphragm; this then allows the diaphragm housing to drop from the base of the siphon as seen in figure 4.15.
8 With the old washer removed a replacement can be inserted. These can be purchased, if you are lucky to find one of the same size; however, I personally have always used thick plastic polythene sheeting and cut out my own, simply laying the old washer on the plastic as a template. The type of plastic you require is the type used as a damp proof membrane or one of those heavy-duty plastic builder's bags. When you get the old washer out you will see the type of plastic I mean.
9 With the new washer cut, everything is replaced in the reverse order. All jointing washers should be in a good condition, but where they have perished simply wrap some PTFE tape around the joining parts (not around the threads) where the old jointing washer or material was.
10 Turn on the supply and test to see if it works. It's hoped that this has been another job well done!

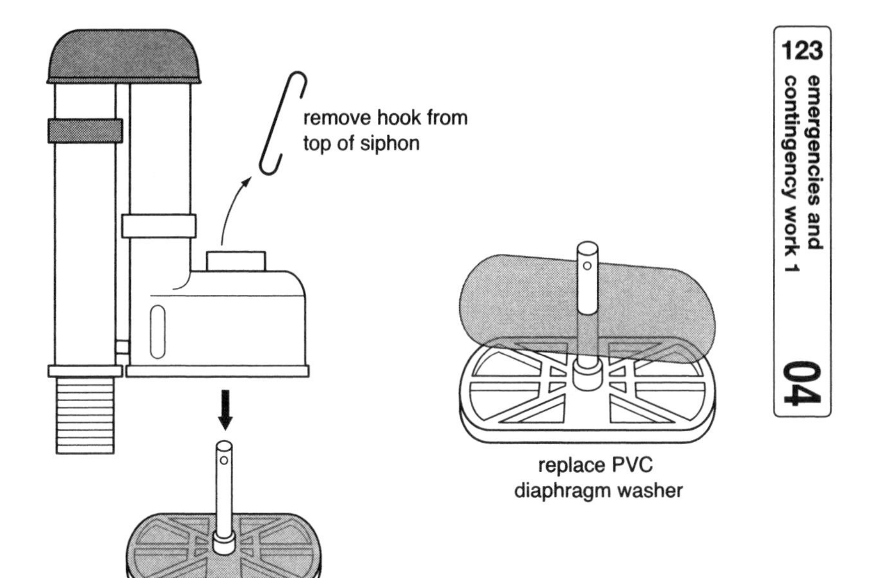
remove hook from
top of siphon
replace PVC
diaphragm washer

figure 4.15 replacement of WC siphon washer

Where a close-coupled toilet cistern is encountered there is, unfortunately, a little extra work to undertake before the siphon can be removed from the cistern. The cistern has to be physically removed from its location bolted to the WC pan. To do this:

1 Turn off the water supply to the cistern and undo the pipe connection to the float-operated valve.
2 The overflow connection will also need to be disconnected. If this is at the bottom of the cistern only undo the nut that connects to the pipe going outside and do not entirely remove the internal plastic tube from the cistern, otherwise the water in the cistern will drain onto the floor.
3 Next, remove the two screws holding the cistern back to the wall.
4 Finally, remove the two wing nuts found beneath the cistern, one on either side of the back of the WC pan, holding the cistern down tight onto the pan itself. The cistern is now free to move and can be lifted from the pan.
5 Tip out the water from the still full cistern into the WC pan.

6 With the cistern removed you will see a big black foam washer pushed over the securing nut of the siphon, often referred to as a donut washer. Simply pull this off and replace it with a new one (which you can get from a plumber's merchant) when re-assembling the WC upon completion of the repair.
7 Follow the procedure described above to remove the siphon and replace the washer.
8 Finally, reassemble the components in the reverse order. In the unlikely event that you cannot obtain a new donut washer it is possible to apply a large ring of 'plumber's mait' (see **appendix 2: glossary**) as an alternative. However, if you use this it is essential that the pan and cistern connecting parts are absolutely dry, otherwise the plumber's mait will not form a suitable seal.
9 Turn on the supply and test to see if it works.

Valved type

These flushing devices have only been installed since the turn of the twenty-first century and therefore are relatively new in the scheme of things (see figure 1.8). When you operate the push-button mechanism to flush the cistern the valve inside should be lifted from its seating to allow the water to discharge directly into the cistern outlet. If the unit fails to flush it is generally due to a broken component and, in most cases, the whole internal flushing valve will have to be replaced as spares for these devices are not generally available off-the-shelf.

If you are lucky you may be able to purchase an identical unit, making a replacement a relatively simple process. Looking at the new component you will notice that there is a facility to turn and remove the valve from its base plate. So, once you have done this and removed the existing valve unit within the cistern, the damaged part can be replaced without the need to remove the cistern.

Remember that you will need to turn off the water supply before carrying out this task.

Toilet continually discharging water into the pan

There are reasons that this problem might occur, and generally it will be the result of one of the following.

Damaged or split siphon or flushing valve

In this case you will need to replace the flushing mechanism in its entirety. In order to do this, follow the guidance for a siphon type of flushing mechanism above but, instead of replacing the washer, replace the whole internal flushing component.

Worn out washer (valved type)

If a replacement washer for a valved flushing cistern is available (it will depend on the manufacturer), this should be your first course of action but, alas, these are not generally available and the entire mechanism may need to be replaced.

Grit accumulated beneath the valve washer (valved type)

Where grit is preventing the valve fully dropping to seal the outlet you will have to twist the valve anticlockwise to release it from its base plate, at which point you can inspect it. When doing this you will need to turn off the water supply to the cistern. If the washer is damaged the valve section may need to be replaced.

Siphonic action failing to stop (siphon type)

Where the cistern continues to flow due to continued siphonage it may be that the cistern is filling too rapidly, in which case close down the isolation valve a little. Alternatively, it might be that the piston is not dropping once the lever arm has been operated. In this case you will need to investigate to find out what is blocking this action.

Water discharge through an internal overflow

This means that the float-operated valve is not operating correctly and is not closing off the supply, in which case you should refer to the notes below relating to the toilet cistern overflowing.

Toilet or storage cistern overflowing

Should you find that water is dripping or pouring from an overflow pipe outside your building, a cistern is probably overflowing due to the float-operated valve (ballvalve) failing to close off the water supply. There are several possible causes of the valve malfunction, including:

- a faulty washer – this is the most likely cause; the washer simply wears out and perishes over time
- limescale – causing the components to rub tightly together, preventing it from moving freely and closing
- the float itself may have developed a leak and have filled with water, making it ineffective, but this is quite rare – if this is the case simply replace the float.

If you call out a plumber to make the repair they are likely replace the entire valve. Plumbers today often do not repair ballvalves as a new one only costs around £5–£8 and it would take longer to repair it than to replace it. Replacing the valve also allows them to offer a better guarantee for their work.

Replacing the valve is quite a simple process and is completed as follows:

1 Turn off the water.
2 If you have a storage cistern, lower the water level by flushing the WC or opening a tap.
3 Remove the old valve (see figure 4.16). Usually it is possible to undo the large union nut inside the cistern which allows the valve to come away for servicing purposes. The bit that is left is that which holds the valve in the cistern and onto which the water supply connection is made. You can now simply undo this nut on the new valve. If there is no internal union nut you will need to replace the entire valve.
4 Install the replacement valve.
5 Adjust the water level as required. This is generally indicated by a mark inside the WC cistern saying 'water level', or in the case of the storage cistern is generally 25 mm below the point where it would ultimately overflow.

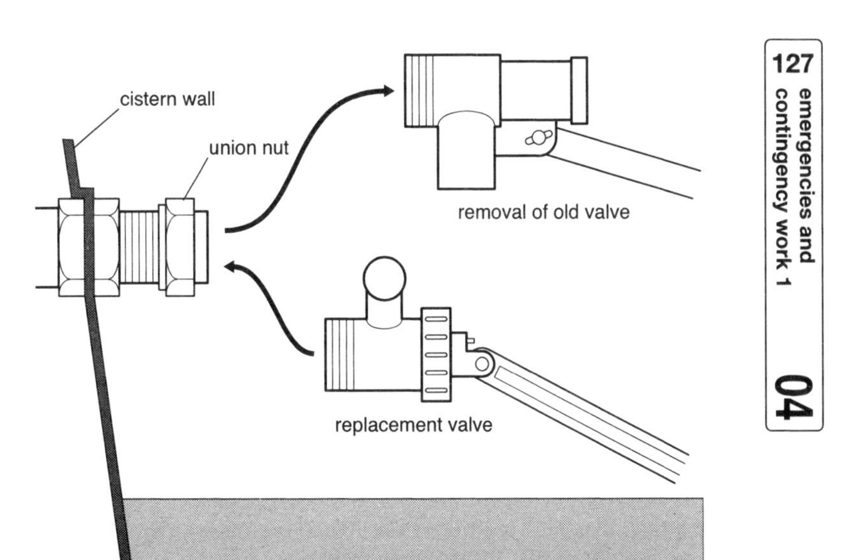

figure 4.16 replacement of float-operated valve

The float-operated valve can, however, be serviced and repaired. This basically requires you to close off the supply to the cistern and to remove the valve as above. Then you simply strip down the component as necessary, cleaning off any limescale or abrasions and replacing it with a new washer.

Float-operated or ballvalve washers are readily available from plumber's merchants, but it should be noted that they are available in a couple of designs: diaphragm ballvalve washers and Portsmouth ballvalve washers. As a consequence you may need to take your old washer into the shop to ensure that you get the right one.

Sometimes the cause of an overflowing cistern is a small piece of grit that has travelled through the pipe and blocked the small inlet hole through which the water needs to pass.

The location of the washer can be seen in figure 1.6. If you have decided to service and replace the washer to the Portsmouth design of float-operated valve (figure 1.6) you will notice the washer is housed inside a small piston. To remove the old washer you simply position a flat-bladed screwdriver in the slot of the piston when removed from the valve and use a large pair of pliers or toothed wrench to unwind the end of the housing, thus exposing the washer. If you do not have a replacement washer, you can sometimes get away with just turning the old washer around.

Toilet leaks when it is flushed

This is a common problem and in most cases it can be fixed by remaking a connection to a component that has worked loose, often by some unknown movement of the appliance.

Where could it leak from? This is the first thing to find out and to do this you simply need to flush the WC and look and feel for the water escaping. Do this as many times as necessary, as it is quite common to think the leak is at one place only to discover later that it is higher up and the water is running down, hidden from view. Possible locations include:

- in a flush pipe joint, where a low- or high-level cistern is used
- at the point where the close-coupled cistern sits on the pan
- at a crack in the porcelain pan itself
- at the outlet connection to the drainage pipe.

With the exception of the cracked pan, which clearly would need replacing, all of the above can be repaired as follows.

Leaky flush pipe joint

This would be a leak from the flush pipe either as it leaves the cistern or as it adjoins the pan.

When the leak occurs as the pipe leaves the cistern

1 The first and simplest thing to do is to try and tighten the large nut (turning it clockwise) that holds the flush pipe to the threaded connection of the siphon as it leaves the cistern base. If there are two nuts *do not* turn the big nut holding the siphon into the cistern. If tightening the nut does not work you will need to unwind it and look at the jointing material beneath. No water will come out when you disconnect this, as water is only present during the flushing operation.
2 With the nut unwound, a rubber ring that has been forced into the joint making up the space between the flush pipe and the siphon will usually be found. In most cases you can apply a few turns of PTFE tape around the existing ring to give it

that additional volume to fill the gap. Do not wind the PTFE around the thread of the siphon as this will do nothing and may in fact prevent you from making a sound joint. The joint is formed where the jointing material is forced into the gap.

When the leak is where the flush pipe adjoins the pan

1 In this case it is likely that you will need a new flush pipe cone or connector. To replace this you may need to undo the cistern connection end of the flush pipe, as identified above, to give you some additional movement, otherwise you simply pull the flush pipe back from the pan, possibly turning it to the side if room is restricted. The joint is only a push-fit type joint, although there are a few different designs, such as those shown in figure 4.17.

2 Once you have removed the old material or connector you can replace it with a new flush pipe connector, replacing everything in the reverse order. You may experience a little difficulty in pushing the flush pipe into the joint when using the insert cone type. This can be facilitated by the use of a little lubricant, such as washing up liquid. The order of assembly for this type of joint is first to place the cone inside the inlet horn of the WC, then to push the flush pipe into the cone.

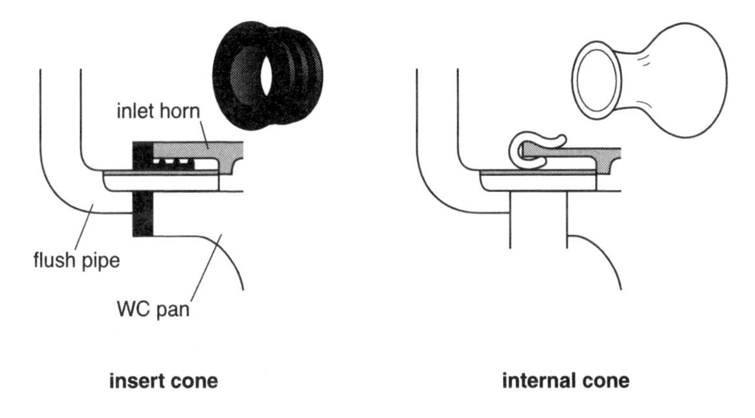

figure 4.17 flush pipe cones

When the leak is where the close-coupled cistern sits on the pan

When water seeps from the space between the cistern and pan when flushed it indicates that the 'donut washer' located over the siphon-securing back nut has perished. The only thing that can be done is to replace this washer. Remove the close-couple cistern (as described earlier) to identify the problem and effect a repair.

Leak from the outlet connection where the pan adjoins the drainage pipe

For well over 35 years now the WC pan outlet connection to the drain has been made using a flexible plastic connector, which either forms part of a plastic drainage pipe or is a device such as a 'Multiquick', which is a patented WC pan connector (see figure 8.7).

These flexible joints are very durable yet, like everything, are subject to possible damage. When this joint is leaking, the best course of action is probably to replace it with a new flexible pan connection. In order to do this you will need to remove the WC pan. Where a low- or high-level cistern with a flush pipe has been used, you will not need to turn off the water supply and remove the cistern, but if a close-coupled pan has been used you will need to remove the whole lot in order to remake the joint.

Where older cement jointed connections have been made, such as in securing the pan to the floor or in forming the outlet joint itself, you may find that the pan cannot be removed and your only hope is to apply some form of sealant, such as silicone, over the crack in the joint, but in truth, the days of the pan may be numbered.

For more advice on removal and replacement of the pan see Chapter 08.

In this chapter you will learn:

- how to prevent noises within your pipework
- how to solve problems with the hot water system
- how to solve problems with the central heating system
- how to unblock your drainage pipes.

Continuing from Chapter 04, this chapter aims to look at more of the tasks that you may need to carry out in the event of something going wrong with your plumbing system.

Burst pipes

The uncontrollable discharge of water from a pipe rapidly gets your heart racing. This is where dividends are paid for your vigilance in locating and isolating the necessary stop valves applicable for each part of the various systems (as discussed in Chapter 04). If you have not already done so, now might be the time to review the section that deals with turning off the water supply.

Where water begins to accumulate above a plasterboard ceiling, the ceiling often begins to bulge. If this happens it is always advisable to make a small hole at the lowest point of the bulge, thereby letting the water out, which can be caught into a bucket. Failure to do this may eventually lead to large sections of the ceiling coming down and creating a great deal of mess and damage. Making a water release hole can also prevent water accumulating above the ceiling and running onto electrical equipment, causing additional problems.

If the water leak is the result of someone banging a nail into a pipe, the easiest way to minimize the water flow is to pop the nail back into the hole made in the pipe. It will probably continue to leak but the nail will greatly stem the flow while you drain down the system via a suitable drain-off cock.

If, for some unknown reason, you cannot isolate the water supply, you could get a hammer and flatten the relevant pipe section; this is not guaranteed to stem the flow but provides a little hope in a desperate situation. Remember: turning off the water mains supply inlet stopcock found at the entry to the property will eventually cause all water to stop flowing. Another option is to turn off every stopcock or valve that can be found. However, here are some indicators that will provide clues as to what system is leaking:

- Can you hear the float-operated valve running in the loft storage cistern? If so, the leak is not on the mains supply.
- If you turn off the mains supply only, does the leak stop immediately? If so, this indicates that it is fed directly from the mains supply.

- Is the water hot, suggesting that it is from the hot water or central heating system? Where the leak is from the hot water or central heating system it is advisable also to turn off the heat source.

Finding the leak

Once you have stopped the water flow you can begin to control the situation. If the leak is in a section of pipework that is hidden from view, such as above a ceiling, the first thing you should do is expose the pipework where the leak was most apparent by lifting the floorboards or removing any covering panels. Now turn the water back on for a short while in order to pinpoint the leak. Don't be surprised if, when you turn on the water, the leak is not from the area you suspected. Water has an uncanny way of travelling long distances undetected.

When you turn the water back on, consider again the clues above which may give some indication of what system is leaking. If you hear cisterns filling in the loft, look to see which cistern is filling. If it is the f & e cistern you know the heating system is leaking. If it is the larger cold storage cistern it will be the low-pressure hot or cold water that has the leak. Each of the cold water outlets from the cistern can be closed off to pinpoint which pipe is leaking.

A bit of detective work often needs to be done to locate the source of a leak. You will need to call on your understanding of the system designs identified in Chapters 01 and 02. You may need to expose more pipework and listen very carefully to the sound of the water hissing from the pipe.

One of the most difficult leaks to locate is one beneath a sand and cement floor screed. The water seems to travel everywhere through the channels preformed for the pipes, making detection very difficult. It is invariably a case of trial and error, exposing test holes in the floor to find the wettest sections. Eventually, however, the point of discharge will be found, as will the supply isolation valve. The rest is now basic plumbing, cutting out the affected section of pipe and replacing it. For this work see the notes in Chapter 06 about jointing pipework.

Noises from the pipework

Water flowing through pipes and into vessels can cause a variety of noises, all of which are quite annoying in their own way. Sometimes we put up with these noises because of the cost to cure the problem. The key thing is to install the system correctly in the first place and most of the problems will never occur. What kinds of noise might you have?

- one or two loud banging noises, usually when a tap is closed
- a series of rapid banging noises
- a humming sound in the pipework
- a shushing noise as water passes through the pipework
- noise transmission generated by a pump
- creaking floor timbers
- splashing noises as water refills a cistern
- noises from a boiler, like a kettle boiling
- bubbles gurgling up through the pipework
- gurgling noises from an appliance waste trap

This is quite a long list and is not exhaustive, but it represents the more common situations dealt with below.

One or two loud banging noises, usually when a tap is closed

This is the classic water hammer sound. It is the result of a stopcock jumper/washer or non-return valve rapidly closing due to a sudden back surge of water, caused by the rapid closure of a valve or tap. This noise can also be created by pipework that has not been fixed securely, so it flaps about. Securing loose pipework may cure the problem, but if not:

1. Slightly turn down the incoming supply stopcock, reducing the volume of water flow and thereby preventing these back surges.
2. Where water flow cannot be compromised it is possible to purchase a small expansion vessel to take up the shock wave. This expansion vessel is similar to that used for the unvented domestic hot water system, but a lot smaller. There are special devices that can be purchased specifically with this problem in mind.

A series of rapid banging noises or a humming sound in the pipework

I have linked these two together because although they are different sounds they are, in fact, caused by the same thing. It is the sound generated by the float-operated valve in a storage cistern rapidly opening and closing as it rides up and down on the small ripples or waves formed on the surface of the water within the cistern. The waves are formed as water flows into the cistern when the float-operated valve opens to make up the water level due to some being drawn off from the cistern.

Sometimes the plastic cistern has been installed without the metal reinforcing piece that came with it, thus the cistern wall flexes as the float rides the ripples on the water. There are several possible cures to this problem:

1 Secure the float-operated valve (ballvalve) by fully supporting the cistern wall.
2 Replace the normal 100 mm diameter float with a larger ball float.
3 If a larger float cannot be obtained, secure a damper plate to the lever arm to create a larger surface area (see figure 5.1).
4 Fit a baffle within the cistern to prevent waves forming. This is basically a dividing plate to reduce the total surface area of the water.
5 Turn down the incoming supply stopcock to reduce the water flow into the home.

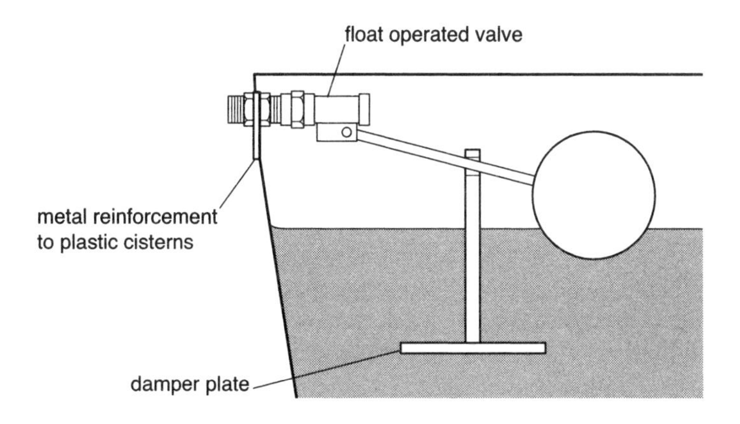

figure 5.1 prevention of ballvalve murmur

A shushing noise as water passes through the pipework

This noise is often generated where the installer has failed to take the small internal burr off from the pipe when using a copper tube cutter. It is also sometimes generated where pipework has been run within a timber stud wall. The plaster boarded timber studwork acts as a resonator, amplifying the sound of the water flowing through the pipe. When running pipes within timber stud walls the pipes should ideally be insulated and the pipe clips located on rubber or felt mountings to stop this transmission of noise.

Curing this problem after the event is often very difficult. Again, try turning down the supply stopcock. It may cure or at least improve the problem. Sometime this noise is generated in central heating pipework, in which case you could try turning down the pump pressure setting.

Noise transmission generated by a pump

Where this problem occurs with a central heating system, generally turning down the setting if a variable speed pump has been installed will alleviate the problem, however, this may create a different problem in large heating systems in that the furthest radiators from the pump may not get warm.

Where the pump noise is generated from a shower booster pump it may be that the pump has not been fitted with flexible connections and onto a flexible mounting and this would need to be provided if necessary. Also check the pump is not touching anything that would act as a sounding box and elevate the noise level.

Creaking floor timbers

This is generally the result of running pipework below timber floors and passing it through the floor joists with notches that are barely large enough, or the pipes themselves have been run touching one another. The noises are the result of the copper pipes expanding and contracting as they heat up and cool down.

When passing copper pipework through notches that have been cut in the joist, ideally a felt pad or piece of carpet underlay should be laid to dampen any movement noise caused by the pipe expanding or contracting. The only option is to lift the floorboards and investigate.

Splashing noise as water refills a cistern

This noise can be eliminated by fitting a polythene collapsible silencer tube. These are often fitted as standard to WC flushing cisterns but are rarely fitted to cold water storage cisterns. Within the loft and inside an insulated cistern the noise is rarely heard, but if the cistern is above your bedroom and in a quiet house, it is the sort of noise that sometimes at night is like Chinese water torture.

If you cannot get a polythene silencer, sometimes fitting an inclined ramp, onto which the water can discharge inside the cistern, eases the problem.

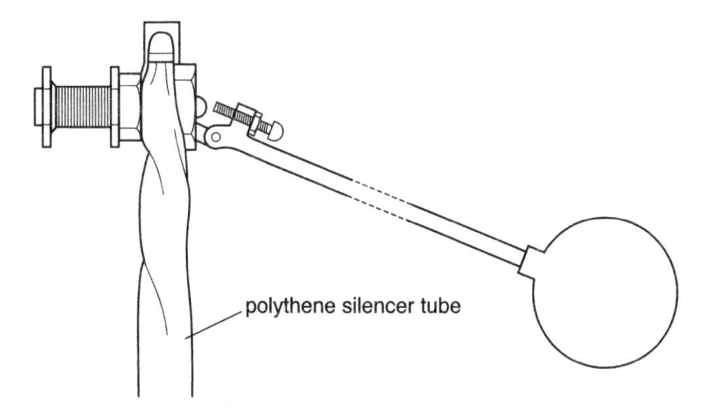

figure 5.2 fitting a polythene silencer tube

A noisy boiler, like a kettle boiling

Noises from the boiler, such as the sound of bubbling water, can be due to several causes. If the system used to work well and the problem has only just started for no apparent reason, it is possible that a narrow pocket of air has become trapped within the boiler, perhaps due to the formation of scale or corrosion. The noise is generated by the formation of steam and its subsequent condensing within this area of trapped air. The only remedy, apart from a new boiler, is to treat the system with a descaling solution. Where a power flush is sought this may require the services of a reputable heating engineer; however, depending on the age of your system and the materials it's made from, e.g. aluminium, copper or steel, several manufacturers produce chemical cleaning solutions, available from any plumber's merchants. These come with the necessary application instructions and can be administered to clean out your system.

Bear in mind that when using these acidic solutions, sludge that has been blocking a corroded radiator or preventing a leaking joint, once removed may expose the fault and leave you with a system that now leaks. But you must remember that they were there already and at least you will find the leaks under a controlled condition and they will not simply spring when you are not at home.

The noise from the boiler, however, may not be the result of trapped air but could be due to impingement of the flame directly onto the heat exchanger within the boiler, causing local hot spots where steam forms and collapses. This requires a specialist heating engineer to make the necessary adjustments to the flame and, where necessary, to investigate the cause further.

Bubbles gurgling up through the pipework

These sounds are to some extent expected in a new system as trapped air slowly releases from the system via the vented pipework. However, if these bubbling noises continue to flow up through the system it suggests a much deeper problem. It is possible that air is being drawn into the system, in particular the heating system due to the incorrect location of a circulation pump (see Chapter 03).

Air being continually drawn into the system increases the speed of corrosion within the system and apart from the noise generated it should still be rectified in order to extend the life of your system.

Gurgling noises from an appliance waste trap

This is the result of water being siphoned out from the trap. This problem has already been discussed in Chapter 01.

Hot water problems

Water getting too hot

If the water is too hot, the most likely reason is that the thermostat on the cylinder is set too high or that the thermostat itself has malfunctioned. Where the temperature is set too high the simple remedy is to adjust the thermostat setting. This needs to be done with a screwdriver. Isolate the power before adjusting an immersion heater thermostat, as you will need to remove the top cover from the unit (see figure 2.7). You won't need to isolate the power if a central heating cylinder thermostat has been strapped to the side of the cylinder.

In both cases the temperature should be set to provide water at 60°C at the top of the cylinder (see Chapter 02). If, however, the thermostat has malfunctioned and simply fails to operate, closing off the heat source, the thermostat probably needs to be replaced. This is a relatively simple process, making the electrical connections with a similar replacement component, but you should isolate the power before doing this. This will be discussed further in Chapter 08.

No hot water or central heating

A lack of hot water or central heating could be due to one of several possibilities. You may have hot water but no heating or vice versa. There is always the possibility of a blockage within the pipeline, such as limescale or sludge build-up, but this type of problem is fairly uncommon. Therefore, the most likely cause and the first thing to investigate is an electrical control fault preventing the power reaching the point where it is required. The following are areas to investigate:

- blown fuse or loss of power supply
- time clock/programmer wrongly set or faulty
- faulty thermostat (room, cylinder or immersion heater thermostat as appropriate)
- faulty motorized valve
- a fault with the boiler or pump.

Electrical faults generally require the assistance of an expert. The engineer will go through the above list and, by a process of elimination, find where the fault lies.

The power supply to the boiler and pump ultimately follows a set route as shown in figure 5.3, and in order to determine the cause of a problem you will need to check that power is going to the first component, then that it leaves that component to move on to the next component, and so on until it reaches the boiler and pump. Along the way you will discover where the interruption in the sequence occurs, so you can focus on the area that is causing the fault. So, for example, if you find that 230 volts is going into the cylinder thermostat yet there is no voltage coming from it, this suggests that this component or the wiring to and from it is at fault.

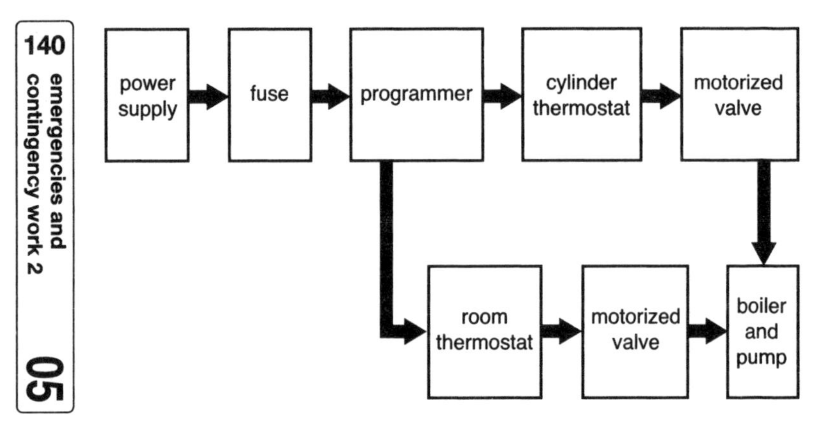

figure 5.3 sequential flow diagram showing power supply to the boiler and pump

Radiators not getting hot

If the radiators fail to get hot this may be the result of an electrical fault as discussed above. However, assuming you have an electrical supply to the pump and boiler it might be a problem with the pump itself. You can check whether the pump impeller is going round simply by placing the end of a large screwdriver up against the pump and putting your ear to the handle. This transmits the sound along the screwdriver shaft to the handle and you will hear if the pump impeller is going round.

You can investigate the operation of the impeller further by removing the large central screw from the body of the pump, out of which a little water will discharge. Behind this big screw you will see another smaller screw head that will be rotating if the pump is in operation. If not, try to turn it with your screwdriver; if you are lucky it will start up and flick from your screwdriver as it rapidly turns. In this case, replace the outer large screw to stop the water seepage. I hardly dare say it, but sometimes giving the pump a tap on its side with a hammer will nudge a pump back into action. If the impeller fails to turn it will need to be replaced.

Replacing the central heating pump

Once you have decided that the pump is faulty, you will need to buy a replacement of a similar design. The task is then completed as follows:

1 Isolate the electrical power supply to the pump and boiler by isolating the circuit and removing the fuse. Once you have

confirmed that the power is dead, remove the old wires where they enter the pump.

2 If you are lucky there will be a water isolation valve on either side of the pump. These are operated by turning the two slotted heads on the valves one-quarter of a turn with a screwdriver or spanner, thereby turning the slots across the pipe instead of in line with it, as they were originally. Where there are no isolation valves, or these are ineffective, you will need to drain down the whole heating system (see Chapter 04).
3 With the water isolated, you can now undo the large nuts on either side of the pump and remove it.
4 With the old pump out, position new sealing washers, if these are used for the mating surfaces, and use a little jointing paste where the components meet as the new pump is inserted.
5 Firmly tightened the joints to secure the new pump in position.
6 Now turn on the water supply and check for leaks.
7 If all is sound you can remake the electrical connections and test the system.
8 Once the new pump is in place, the speed, if it is a variable speed pump, will need to be set to the lowest setting and only increased if all the radiators fail to get hot. If the speed is set too high it might create unacceptable noises within the system.

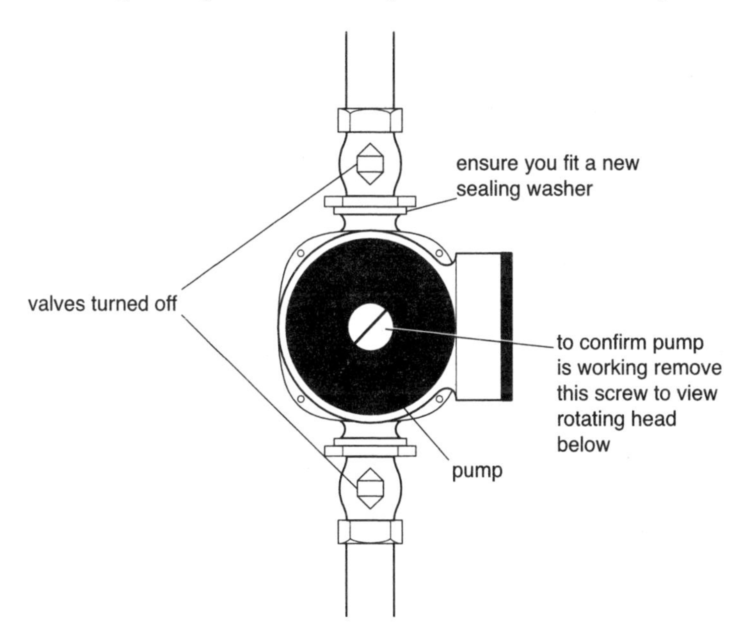
ensure you fit a new sealing washer
valves turned off
to confirm pump is working remove this screw to view rotating head below
pump

figure 5.4 replacing a central heating pump

Radiators aren't getting hot, but the pump is okay

If the radiators fail to get hot but the pump is working, the system may be blocked with sludge caused by corrosion within the system. Should this be the case then you will need to descale the system using a special acidic solution to dissolve it, as discussed earlier.

Sometimes a radiator only gets warm around the sides and along the top and has a cold spot in the middle. This is a classic sign that corrosive sludge has accumulated in that part of the radiator. Again, it may be possible to solve this problem by using a descaling solution. Alternatively, the radiator can be removed and subjected to individual treatment and flushing through with a high-pressure hose.

There may be just one or two radiators on your system that are not getting warm. Assuming that the valves at each end of the radiator are open, the first thing to check is that they are not cold simply because they are full of air. Air is expelled from the system radiators through a small air-release valve located at the top and to one side of the radiator (a process referred to as bleeding). Do this as follows:

1 Turn down the room thermostat. This will turn off the pump. (The reason for turning off the pump while bleeding the radiator is to ensure that air is not sucked into the system if the pump is creating a negative pressure within.)
2 Use a special square-headed radiator key to open the air-release valve, turning it anticlockwise. You will hear the air being forced out and eventually water will appear at this point, whereupon you simply close the air-release valve.
3 Turn the room thermostat back to the desired setting.

If a particular radiator continually accumulates air, this suggests that air is being drawn into the system, possibly due to the incorrect positioning of the circulation pump. This situation must be addressed as the air that is being drawn into the system will speed up the corrosion process inside the system and very soon you will be experiencing leaky radiators that have corroded from the inside. Correct pump location has already been discussed in more detail in Chapter 03.

If some radiators still remain cold, the system might be too large for the pump. A particular pump only generates so much pressure and will only push the water so far, therefore a larger pump may be required. The pump may have variable settings and it might be possible to increase its speed and pressure by making a simple adjustment on the side of the pump itself.

Another possibility is that some of the radiators closer to the pump have their lockshield valves open to such an extent that they are taking all of the circuitry flow of water, in which case they need to be closed a little in order to balance the system (see Chapter 03). A simple test to see if balancing is required is to close off the manual radiator valve operating heads to several radiators that are working fine to see if the cold radiators then get hot; if so, you need to balance the system better or get a more powerful pump.

Leaking radiator valve

Sometimes when a radiator valve is operated it leaks from the nut at the point where the spindle turns. This can only be seen when the plastic head is removed. This leaking joint is often the result of the valve not being used regularly. The simplest cure and often all that is required is to tighten up the gland nut (see figure 3.11). Where this does not cure the problem the gland will need repacking. To do this:

1 First, turn off both radiator valves. To close the lockshield valve you will need to use a small spanner. When you close the lockshield valve, count the number of turns it takes and when required only open the valve by that number of turns.
2 With the valves closed, simply unwind the gland nut, pack a few strands of PTFE around the spindle and push it into the void into which the packing gland nut screws, poking it down with a small screwdriver.
3 Now replace the packing gland nut, tightening it just sufficiently to squeeze the new packing material within the gland.
4 Re-open the radiator valves and test it.

Repacking this gland nut is basically the same procedure as repacking any gland, as described with reference to a leaking stopcock or tap in Chapter 04 (see figure 4.11).

Blockages to the waste water pipework

The most effective weapon used by the homeowner or a plumber when tackling an obstruction is a plunger. The plunger, when applied effectively, causes atmospheric pressure (around 1 bar) to push directly onto the blockage. This will not mean much to most people so using the old British Imperial measurement this equates to 14.7 pounds per square inch. That's like seven bags of sugar applying a force against an area the size of your watch face or, in the case of a 100 mm drainage pipe, 88 bags of sugar pushing onto the blockage. That's a whole lot of force! How does it work? Basically, when the plunger is pulled from the source of the blockage it creates a partial vacuum. It is this partial vacuum or void that the atmosphere pushing down on the earth tries to fill.

Blocked sink, basin or bath

Let's look at the plunger first.

1 You will require a force cup plunger as shown in figure 5.5. These are easily obtainable from most hardware stores or plumber's merchants.
2 Fill the sink with a fair quantity of water.
3 You must now block up the overflow pipe. To do this, take a piece of rag and stuff it hard against the overflow opening. You must make a good seal here in order to be successful in the plunging operation.
4 All you need to do now is push the plunger up and down over the waste pipe outlet several times.

This will invariably clear a blockage; however, blockages to appliances such as sinks and basins are often the result of soap and fatty deposits. Plunging will give some relief to the problem but will not remove all of the fatty debris and will only make a small hole in the blockage which will soon block up again. In this instance the ideal solution is to remove the trap from the appliance for internal inspection.

Where a plastic trap has been fitted to the appliance this is a relatively easy operation, but for older metal traps more force and therefore care will be required to undo the nut/s of the trap. To remove the trap:

1 Empty as much of the water from the appliance as possible, bailing it out into a bucket.
2 Now position a bucket or suitable receptacle beneath the appliance to catch any spillage and remaining water from within the trap itself.
3 If a bottle trap has been fitted you will just need to remove the lower dome-shaped bottom, as shown in figure 5.5. Where a tubular trap has been installed you will need to remove this in its entirety, as follows:
   a) Undo the large nut that joins the trap to the appliance waste outlet turning it anticlockwise.
   b) Now undo the nut that joins the trap to the pipe; this will enable the trap to be removed.
4 Be prepared for a sickly sight of fat, hairs and general grime. However, once all this rubbish has been removed (undoing the third large nut adjoining the two sections of the trap if necessary), you will have a clean trap with an effective internal bore.
5 Before replacing the components, just look into the outlet pipe for any further signs of blockage. If there is excessive blockage it may be time to consider using a series of long drainage springs to poke down the tube, or you could remove the whole pipe section and replace it, but this is not usually necessary.

During this process no water apart from that held within the sink and trap itself will flow from the appliance. In most instances the trap is fairly easy to access, but sometimes where a pedestal basin has been used, you may experience some difficulty in accessing the nut that adjoins to the waste outlet of the appliance. It might be possible to ease the pedestal forward a small amount to gain better access, but take care as it is designed to give support to the basin and is easily chipped, being made of porcelain.

When you reassemble the waste pipe, take a little care when doing up the nuts as they are made of plastic and it is easy to cross-thread a joint, preventing it from doing up tightly. The seal that was in place prior to undoing will probably still be fine to reuse, but if necessary a few turns of PTFE tape to the mating surfaces where a damaged sealing washer was is usually sufficient. Do not wrap PTFE tape around the pipe threads themselves as the nut and thread are just used to pull or clamp the two mating surfaces together, crushing jointing material in place to form the seal, and they do not themselves form the watertight seal.

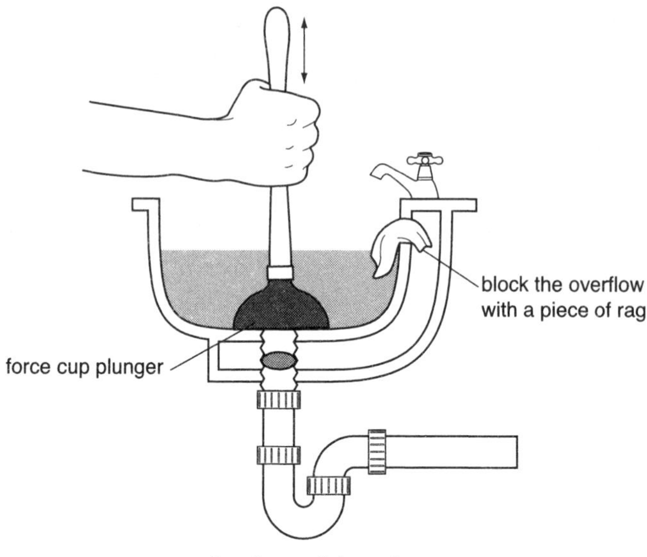

plunging a sink waste

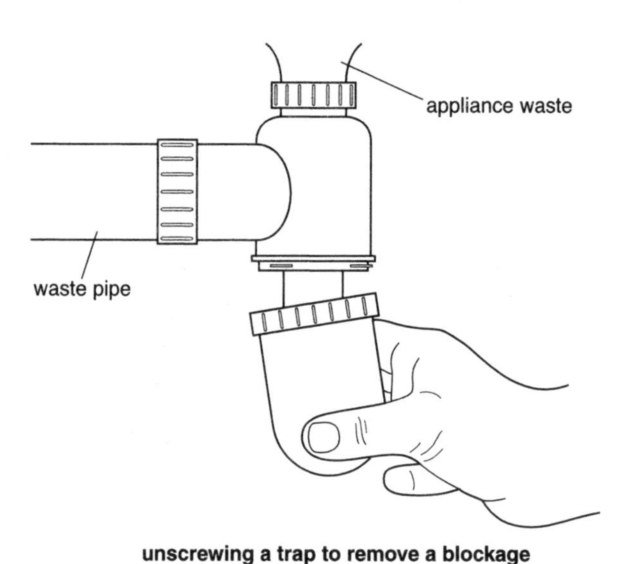

unscrewing a trap to remove a blockage

figure 5.5 unblocking a sink or basin

It is possible to avoid stripping down the trap by purchasing a drain cleaning solution, available at most hardware stores. These can be most effective, using an acid solution to dissolve the offending matter, and this option should not be overlooked.

Blocked toilet

Spending a few pounds on a drain rod and a 150 mm rubber plunger to screw onto its head could easily save you £100s in plumber's call out charges. When someone's toilet blocks the natural instinct is to panic and to wish the problem would go away as quickly as possible. When you flush the toilet the bowl fills with foul water which just sits there. It may slowly drain away but the blockage still remains, and after the next flush the water will back up and fill the bowl again.

If you call out a plumber they will probably fix the problem within 30 seconds of arriving, leaving you happy to pay whatever they ask. There is no magic – it is simply a matter of them using their plunger to create the 1 bar pressure needed to dislodge the blockage. So what do you do?

1 Obtain a drainage rod or chimney sweep's rod that has a thread on one end. Onto this screw a 150 mm (6") drain plunger obtainable from plumber's merchants.
2 Ensure water, however disgusting, is in the WC bowl, or flush the appliance so that it fills and backs up.
3 Push the rubber plunger back and forth down inside the pan, back towards the trap, as shown in figure 5.6.

Hey presto! With any luck the problem is cured. I recall once clearing a blockage using this technique without the use of a plunger. I used an old-fashioned floor mop onto which I secured a plastic bag; this made a suitably effective plunger. Plunging can be very effective. Toilets that remain blocked after plunging could have a blockage further down the pipeline and air is simply getting in via the open vent pipe at the top of the drain, relieving the partial vacuum you are trying to create.

Blockages further along the drainpipe might also affect other appliances, in effect putting several appliances, such as sinks and baths, out of action.

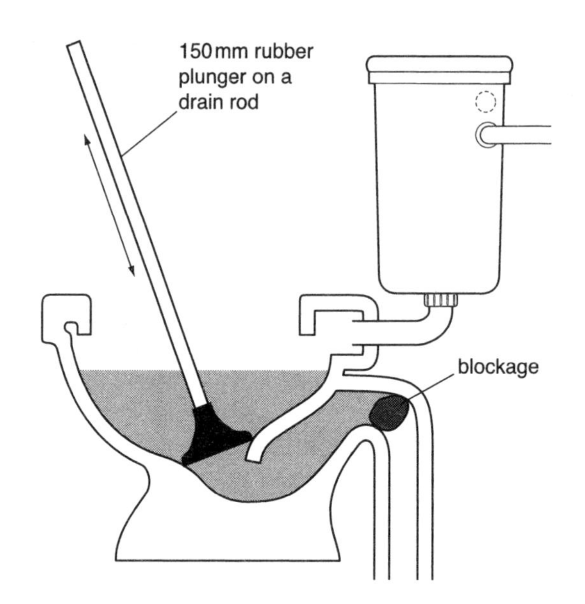

figure 5.6 unblocking a WC pan

Blocked drains

Aaargh! What a nightmare! Nobody likes blocked drains. Everything in the household may be put out of action as a result of this kind of blockage. The first course of action is again to consider the plunger.

Let's assume you lift up the inspection/manhole covers outside your home and find they are filled with sewage. Arm yourself with a set of drainage rods now, not just the single one required to unblock a WC pan (these can be hired out quite cheaply from most hire centres). Secure a 100 mm (4") plunger onto the end and insert this into the next dry manhole chamber down from the blockage, aiming towards the one full of liquid. Inserted the rod several meters and then pull it from the pipe. It is hoped that this will create the suction required to dislodge the blockage. Nine out of ten times it will! If you cannot locate a dry inspection chamber you will need to try and pass your rods, with its plunger attached, through the sewage towards the outlet to pass it into the pipe. Again, once inserted simply push and pull the plunger to create the pressure to dislodge the blockage.

Warning! When the drainage rods are inserted into the drain, never turn your rods anticlockwise or you might unwind the plunger from the end of the rods and leave it behind inside the pipe causing a real problem.

Unfortunately, plunging the pipework does not work in every situation, but always try this first. Where this fails you will need to secure the worm screw attachment onto the end of the drainage rods. These are then passed down through the pipe until the blockage is encountered and then you need to give a few forward blows, hitting directly onto the blockage, trying from both directions if necessary, i.e. from upstream and downstream manholes. Remember, never turn the rods anticlockwise.

Drainage systems installed in buildings prior to the 1960s often incorporated a special intercepting trap at the point where the house drain joined the public sewer. These are no longer installed as they were often the cause of blockages. If one of these is blocked in an older property it will need to be plunged in the same way as a WC pan. However, if the blockage is downstream of the trap there is a stopper which allows rodding access towards the sewer. The stopper should be removed by lifting it from its seating, pulling it up with the attached chain.

For smaller pipes a snakentainer or flexible wire is used, which again is passed into the pipe, continually turning clockwise, to dislodge any obstruction.

Clearly, when working on any of these kind of tasks it is absolutely essential that your take the appropriate hygienic measures and wear rubber gloves to ensure you are not contaminated by germs that may be lurking within the drainage system.

One final point regarding blocked pipes is that if you have to remove bolted-on access covers, particularly those inside the house in the above-ground part of the drainage system, give some thought to what might be behind the access door. These are designed to be watertight, so prior to opening them you do not know what is behind the door. You might find yourself covered in a nasty solution, as the pressure of the backed-up liquid contained within could be quite considerable, spraying the contents some distance from the opening.

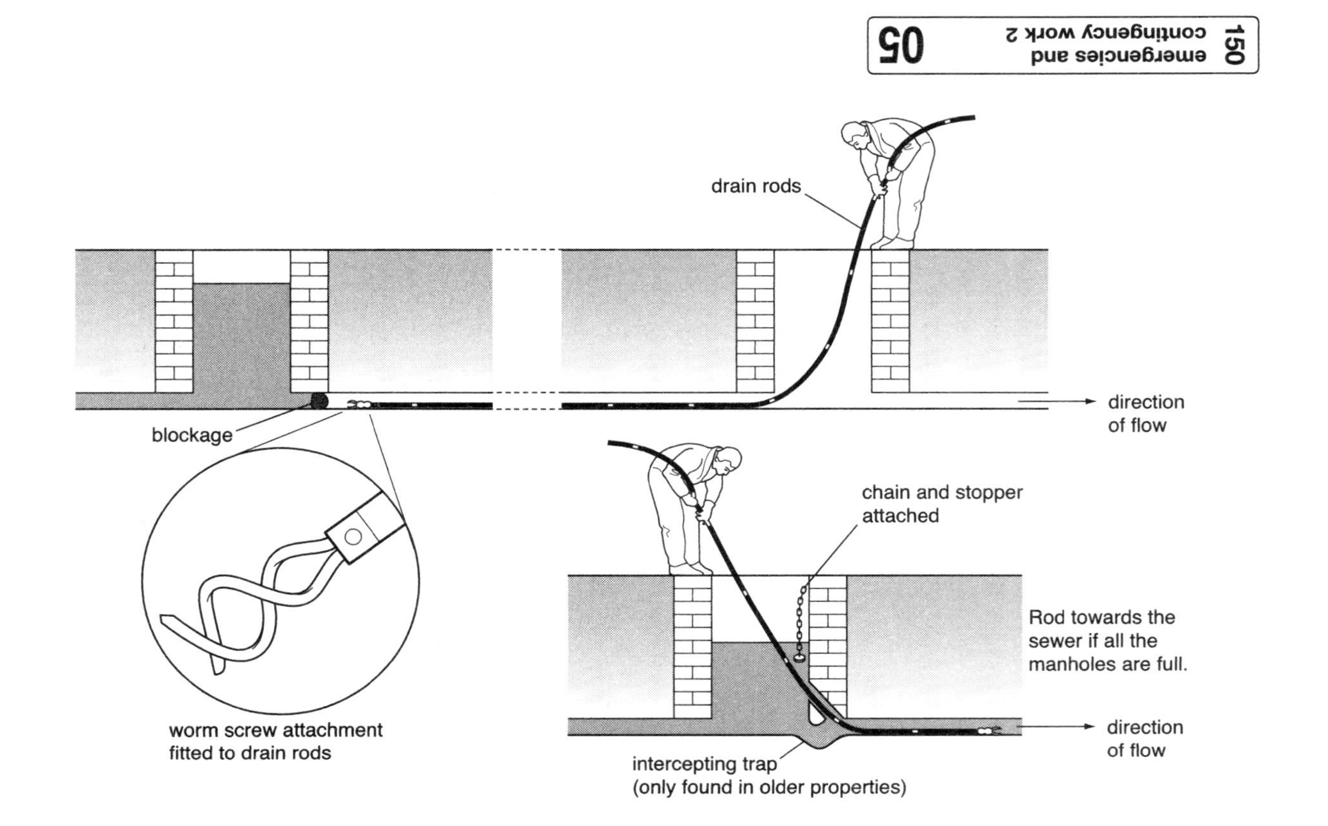
drain rods
direction of flow
blockage
worm screw attachment fitted to drain rods
chain and stopper attached
Rod towards the sewer if all the manholes are full.
direction of flow
intercepting trap (only found in older properties)

figure 5.7 rodding a drain

Blocked gutters and rainwater pipes

Over time there is the possibility that your gutters will collect dust from the atmosphere, plus moss and other debris as it falls from the roof. This inevitably silts up the gutters, making a bedding material for seeds to grow in. Eventually the gutter will overflow as water cannot freely pass to the down pipe. It is very easy to clear out this debris and a useful tool to help with this is a small semicircular section fixed onto the end of a pole, to pull any debris towards you. Simply collect the rubbish into a bucket for removal.

Take care when working up a ladder. If you do not feel confident doing this it may be advisable to call upon someone else to do the work. If you do decide to do it yourself, always ensure the ladder is 'footed' by another competent person, to ensure that it will not slip, and never overreach when working up a ladder. Never lean the ladder up against the gutter itself as it may cause damage, especially where a plastic gutter has been installed, but more importantly your ladder can also easily slip to the side when it is resting against such a smooth surface, making it very dangerous.

Should the rainwater pipe itself become blocked with debris, it poses a much more difficult problem. You can try poking a drainage spring up or down the pipe, but sometimes, especially if it is a plastic rainwater pipe, it is quicker to disconnect the entire pipe section and do the unblocking at ground level. Fortunately it is usually just a blockage at the bottom end of the pipe that is creating the problem, causing the water to back up and come out of the joints, which are not made watertight and incidentally were never intended to be watertight. A blockage at the bottom of the pipe is often the result of a blocked gully, which can simply be emptied physically by hand.

Smell of gas or fumes

Never allow this situation to go unchecked! In the event of smelling a gas leak observe the following guidelines for your own safety as an explosion could result:

1 Turn off all gas appliances.
2 Turn off the emergency gas control valve at the gas meter.
3 Open all windows.

4 Do not operate any light switches and extinguish all naked flames.
5 Call a gas service engineer or your gas supplier.

The telephone number of the national gas emergency service is: 0800 111 999

You should always have easy access to the gas meter. If you have a gas meter box make sure you know where the key is kept in case of an emergency. You can close off the supply completely by turning the handle attached to the emergency control valve just one quarter turn.

The products of combustion are highly dangerous, not just from gas appliances but from any fuel-burning appliances, such as those burning oil and solid fuel. You may smell products other than gas and there is a good chance that these will contain carbon monoxide, which is a highly toxic gas (see Chapter 02). As above, if you smell any fumes:

1 Turn off the gas appliance.
2 Open the windows where the fumes are discovered.
3 If you feel drowsy, evacuate the building to get some fresh air.
4 Call a doctor if necessary.
5 Call a gas service engineer.

In this chapter you will learn:

- about the various plumbing materials and pipes
- about jointing to pipes
- about bending copper pipes
- about specialist plumbing tools
- about working practices
- about concealing pipework.

This chapter aims to look at some of the practical skills that need to be mastered in order for you to undertake some specific plumbing tasks. Often it is a case of knowing how to undertake a job and what particular tools are available to help complete the task. The chapter also takes a closer look at corrosion, what it is and why it needs to be taken into consideration in the way the various plumbing systems are designed.

Corrosion

Corrosion is a chemical attack upon the metal that brings about its destruction. There are various forms of corrosion, of which two types are identified here:

- atmospheric corrosion
- electrolytic corrosion.

Atmospheric corrosion

Everyone has seen atmospheric corrosion: leave a tin can in the garden and within a very short period it will be rusty and full of holes. It is the water and oxygen in the air that causes this corrosion: their presence on the exposed surface of iron causes oxidation. The resultant iron oxide is not stable and falls away, exposing more fresh metal, and the process continues until there is none of the iron left and only a collection of iron oxide, or rust as it is commonly known, scattered on the floor.

Atmospheric corrosion attacks all metals in this way but unless the metal is ferrous (i.e. contains iron) the corrosion formed on the surface of the metal is stable and so prevents any further corrosion. This can be seen by looking at a copper roof that has turned green – the green is the oxidized copper that has formed due to corrosion over a period of time. Copper pipe is unaffected by atmospheric corrosion so can therefore be used for water supplies without fear. If we used iron pipes for water services they would only last a very short time. You may find iron pipework in your home but the iron has been covered with a coating of zinc, referred to as galvanized, so the metal is in fact protected to some degree against atmospheric corrosion.

As discussed in Chapter 03, steel radiators are used in central heating systems and last for many years without rusting. This may seem strange as it is totally immersed in water, but for corrosion to occur there also needs to be oxygen present. There is a fair proportion of oxygen within a sample of water, yet the radiators do not rust because the water is never changed, except for repair work, and within a week or so of filling the system oxygen within it will have escaped back into the atmosphere. So with no oxygen there is no rusting.

Electrolytic corrosion

Galvanized mild steel, although no longer installed within the domestic home, can still be found. This pipework, although protected against atmospheric corrosion, is subject to another form of corrosion brought about by a process known as electrolysis. This is where one metal attacks and destroys another metal lower down the electromotive series. The electromotive series is a list of metals with different abilities to resist destruction by another metal – the metals lower down the list are less able to resist than those higher up the list. Where there is a mix of different metals within a system the metal lowest on the list is totally destroyed first, before electrolytic corrosion begins on the metal next highest in the list.

Electromotive series of typical plumbing metals:

copper
lead
tin
iron
zinc
aluminium

Galvanized mild steel is iron pipe with a coating of zinc. The zinc coating not only protects the iron against atmospheric corrosion but also provides a sacrificial metal to be destroyed before the iron when mixed with materials such as copper. If you look at the list above you will see that the copper would destroy the zinc before the iron is attacked, as the zinc is lower down the list.

Pipework used for water supplies

There is a whole range of fittings that can be purchased which are designed to make all kinds of connections, such as for joining pipe to pipe or a pipe to an appliance. Today you can pick up special flexible pipe fittings designed to assist in making connections to basins, baths and other appliances. In the past these connections could only be achieved with the use of a bending machine. For the last 50 years or so, copper pipe has been the material most widely used for running water through; however, more and more plastics are being used today and depending upon the age of your property other materials may be encountered, including mild steel and lead pipework which would no longer be installed but to which connections can be made.

Connections to lead pipes

Lead pipework installations should always be removed where possible as the material is toxic and can contaminate the water supply. Any connection to this pipe should therefore be made only as a last resort. This might occur, for example, where you need to make a connection onto an existing lead mains cold water inlet supply pipe. The connection is made using a special compression fitting. These joints are similar to the compression joint used for copper pipes (see page 159), only they are much larger and have a rubber compression ring instead of the brass ring used for copper. These compression fittings (e.g. Lead-loc) can be obtained from a reputable plumber's merchant. The replacement of old lead mains should be considered at the earliest possible opportunity, for which a local government grant may be available.

Connections to mild steel pipes

As with lead pipes, mild steel pipes should generally be removed whenever possible, as they have now well exceeded their expected lifespan. It is more than likely that they are excessively corroded inside and are affecting the volume of flow that you should expect. In fact, old steel pipework is one of the major causes of blockages to existing systems of water supply. Connections to mild steel pipework with copper will create additional electrolytic corrosion problems, as discussed above, and if a connection is made it could be the source of a blockage problem within a few years.

Connection to the pipe can be achieved with a similar compression joint to that used on lead pipe, with a rubber compression ring. However, the best joint to use would be to make a connection onto a threaded joint. This would be either:

- a male iron thread (an external thread)
- or female iron thread (an internal thread).

These fittings can be seen in figure 6.1 The threaded connection is made by firstly applying a few turns, in a clockwise direction, of the jointing tape referred to as PTFE onto the male thread of one fitting. This is then wound into the female thread of the other fitting thereby forming a sound bonded joint. The copper or plastic pipe is then made onto this fitting as a compression connection (see below). It is possible to use a jointing paste instead of PTFE tape, but you will need to ensure that it is acceptable for use with the contents of the pipe, as indicated on the side of the tin.

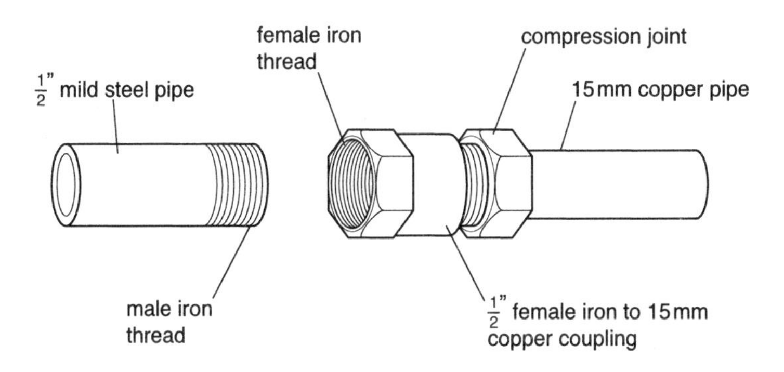

figure 6.1 threaded joints

PTFE

PTFE is the abbreviation for Polytetrafluoroethylene, which is a white coloured plastic tape that is used extensively for making joints to threaded pipe connections, or as a packing material where some make-up to a small void is required. It is readily available and can be purchased at all plumbing merchants. In the United States PTFE is known as Teflon.

Copper pipework and fittings

Copper has become well established as a piping material suitable for all water supplies and in all circumstances. Making a sound water connection to a pipe is a relatively simple operation and once you've mastered this skill you will be able to undertake fairly substantial projects. There are three basic jointing methods used in the home:

- compression joints
- solder joints
- push-fit joints.

Compressions joints

These are made using a fitting that clamps a compression ring onto the pipe and wedges it into the fitting at the same time. To complete a sound joint:

1. First, push the nut onto the pipe.
2. Then push on the brass compression ring.
3. Next, insert the end of the pipe fully into the fitting making sure it reaches the stop.
4. Push the compression ring along the pipe to the mouth of the fitting.
5. Now wind the nut onto the thread of the fitting in a clockwise direction. This pulls the compression ring into the fitting. It is essential that the compression nut is not tightened up too much as this will distort the compression ring inside, which may cause a leak. The joint should only be tightened sufficiently to hold the connection firm. When the water is turned on it can always be tightened a little more if necessary, but once tightened too much no further tightening would cure the leak.

Note that no jointing materials are necessary to make this connection; it is a dry jointing method. However, a trick sometimes used by plumbers, especially where the compression ring used is not new, is either to wrap a ring of PTFE tape over the compression ring or apply a little jointing paste onto the ring to make up for any blemishes. This is *not* applied onto the threads of the fitting, as these are just used to pull the joint together and do not make the seal.

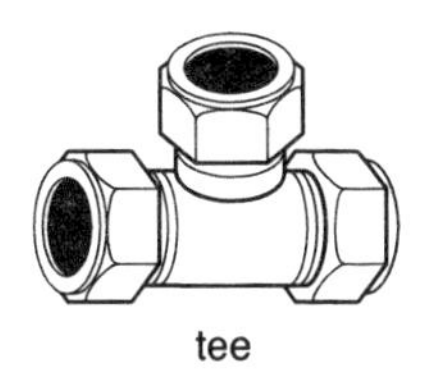

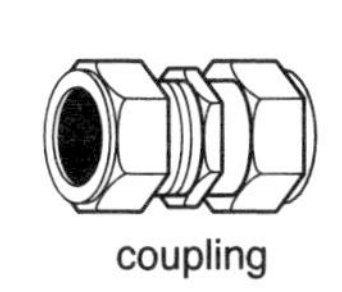

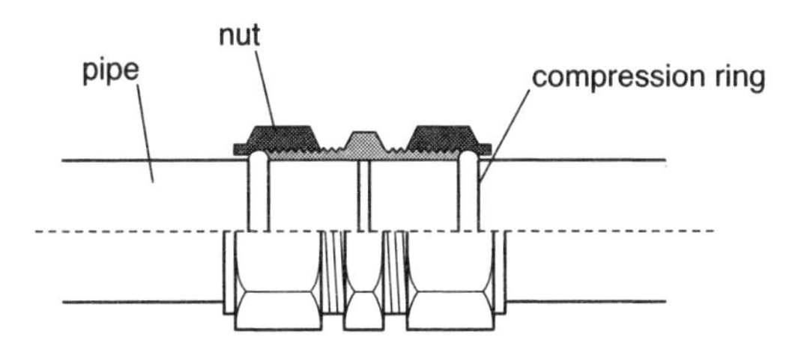

figure 6.2 compression fittings

Soldered joints

These are joints that have been made with the use of a blowlamp, although there are electric soldering machines today that can be used to supply sufficient heat to the joint without the need of a blowlamp.

There are two types of solder fitting: those that contain a ring of solder (referred to as solder ring fittings) and those that require the solder to be applied from a reel (referred to as end-feed fittings). When using the solder ring fittings no additional solder needs to be applied to the joint.

Note that the solder used for hot and cold water supplies needs to be lead free in order to avoid contaminating the water. However, where central heating pipework is being installed it makes no difference what kind of solder you use. Both of these solders are readily available from the plumbing suppliers.

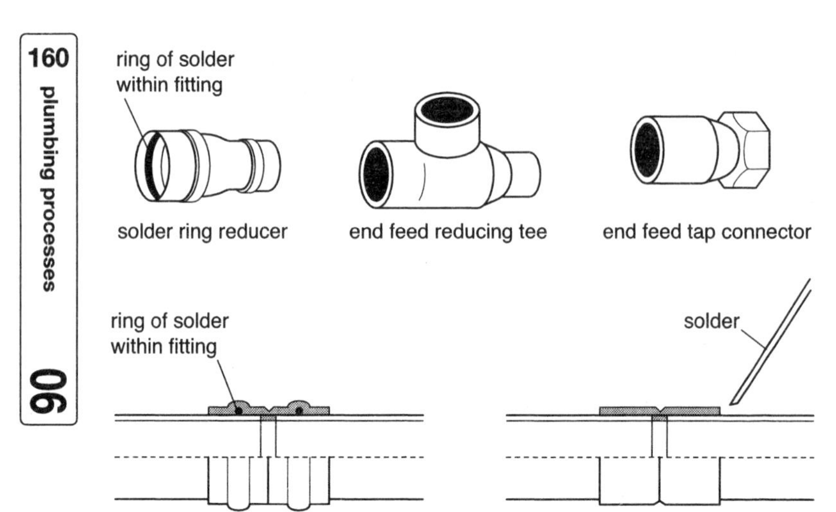

figure 6.3 soldered joints

Soldered joints are made as follows:

1 Adequately clean the mating surfaces of the pipe and inside of the fitting. Do this with wire wool or a special nylon cleaning pad available from plumber's merchants.
2 Apply a suitable flux to the cleaned surfaces. This is a special paste, readily available from plumber's merchants, applied in order to keep the work area clean while soldering, thereby allowing the molten solder to stick to the copper and flow easily. Note that solder will not adhere to dirty or oily surfaces. (There are self-cleansing fluxes that will clean the pipe and fitting when the heat is applied, but take care as they can be aggressive and any residual flux needs to be fully flushed from both inside and outside the pipe.)
3 Ensure there is absolutely no water in the pipe when soldering, otherwise it will not reach a high enough temperature – even the smallest drop of water will prevent the solder from melting.
4 Using a blowlamp or a soldering machine, apply heat to the assembled joint to melt the solder. Apply the solder as soon as it melts – do not simply hold the blowlamp there and burn

away all the flux. In a case where solder ring fittings have been used, the solder will be seen emerging at the mouth of the fitting. Then remove the heat source.

5 Take care not to set fire to combustible materials in the vicinity.
6 Allow the joint to cool before moving it.
7 Finally, wipe off any residual flux – it will make the pipe go green due to its effect of corrosion on the pipe.

Should the joint leak when you test it, you will need to completely remove it and form a new joint, using a new fitting. The problem is most likely to have been a dirty joint. Cleanliness and the application of a flux are essential in order to solder a joint successfully.

Push-fit joints

There is a whole range of push-fit joints available. These joints are very effective and you should not worry that they will not hold the water pressure, as long as you have assembled the joint correctly, inserting it fully into the fitting and ensuring that it is pushed all the way up to the internal stop. The joint is achieved by the use of an internal 'O' ring. When elbow or bend joints are used they have the advantage that they can be swivelled around to any direction, even when water is in the pipe. Because of this freedom of movement, the pipework does need to be fully supported with pipe clips (see below).

Push-fit joints cannot readily be pulled from the pipe as there is an internal grab ring preventing withdrawal. However, they can be dismantled and reused. To remove the joint illustrated in figure 6.4, push the end collet tightly into the fitting and, while holding it close to the fitting, the pipe is pulled out. Different manufacturers use different methods to disassemble the joint and further advice would need to be sought from the manufacturer of a particular fitting if necessary.

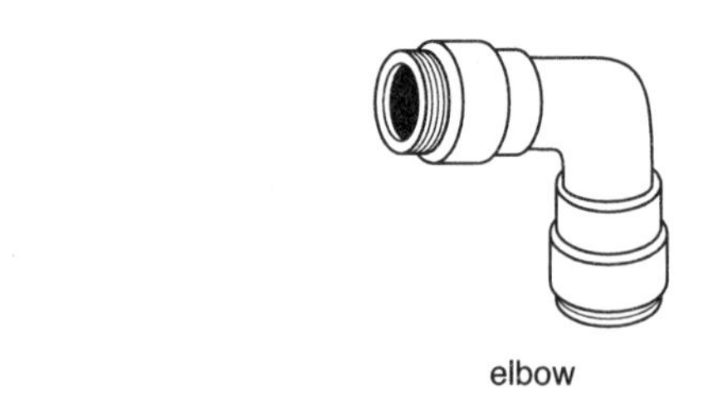

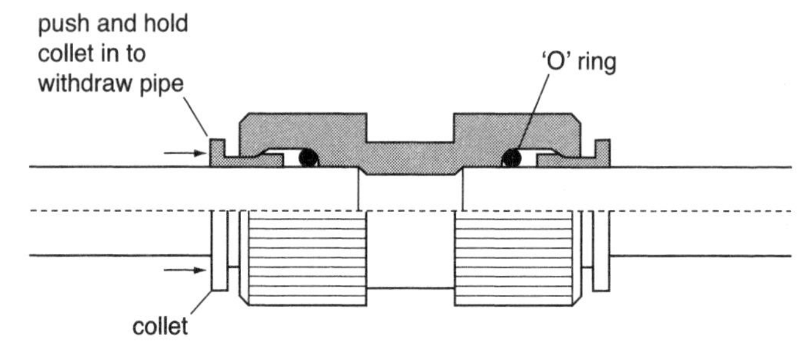

figure 6.4 push-fit joints

Pipe clips

All pipework will need to be securely supported and held firmly by a pipe clip. These should be securely fixed to the wall or adjoining surface at a distance not exceeding the dimensions listed below.

table 6.1 Maximum pipe support spacing in metres

pipe size mm	copper clips horizontal	vertical	plastic clips horizontal	vertical
15	1.2	1.8	0.6	1.2
22	1.8	2.4	0.7	1.4
28	1.8	2.4	0.8	1.5

Bending copper tube

Copper tube can be installed using fittings throughout, thereby avoiding the need to pull any pipe bends. However, this would:

- increase the installation time
- add to the cost of the job
- increase the likelihood of leaks
- reduce the pressure available at the outlet, due to the increased frictional resistance caused by the fittings installed.

It is possible to purchase special flexible pipes and these do have a use in areas such as making the final connections to bath taps, but these would again increase the cost of the work if used extensively throughout the plumbing project and they would look unsightly. Bends pulled directly onto the pipe are preferable, but in order to form a bend you will need either a bending spring or a bending machine.

Using a bending spring

This is the cheap option for pipe bending. The bending spring will only cost around £3–£4 and, if used correctly, will generally be more than adequate for occasional use. However, they are easily damaged and can get stuck inside the pipe if wrongly used. There are several tricks to successfully using the bending spring:

- Don't pull the bend too sharply, i.e. have a long radius to the bend. As a general guide for a 90° bend on 15 mm copper pipe the radius will be something like that if pulled around a pipe of 150 mm (6") diameter.
- Always slightly over pull your bend, then open the bend out again. This will release the spring inside the pipe to assist removal.
- When withdrawing the spring after pulling the bend *do not* just pull hard at the end of the spring, but turn it in a clockwise direction. This is done with the aid of a screwdriver passed into the loop at its end. This tightens the spring up, creating a smaller diameter. Just pulling hard to remove the spring will damage it.
- If it gets stuck, try gently closing and opening the bend a little to free up the spring.

Bending springs are ideal where small bends and direction changes are required and used to form offsets in the pipework. With the spring inserted into the pipe, pull it around a round object or around your knee, keeping the radius smooth and not too sharp. If at any time a ripple begins to form in the bend immediately stop the process and withdraw the spring as it will undoubtedly get stuck inside the pipe.

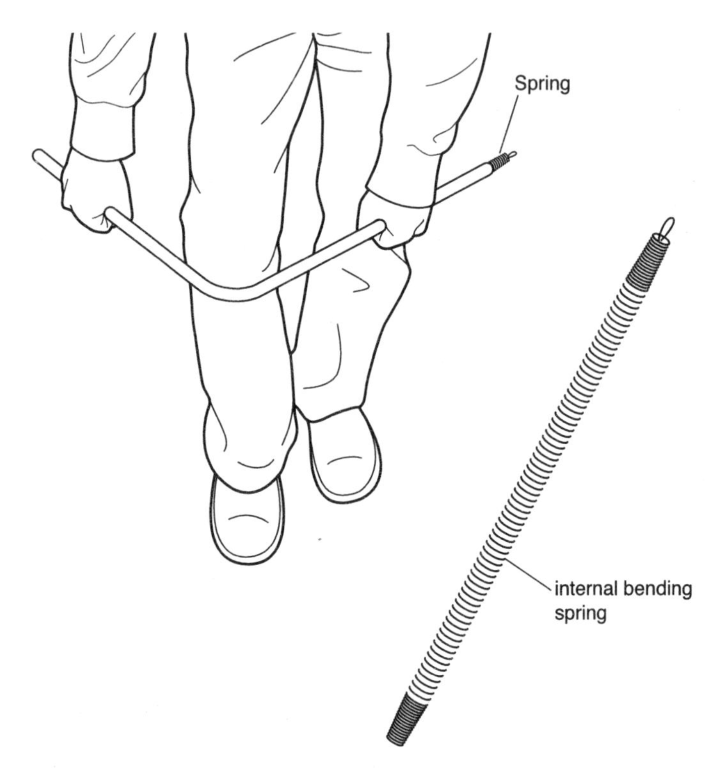

figure 6.5 pipe bending with a spring

Using a bending machine

A small handheld bending machine will last a lifetime, but will cost around £50 or so. They can also be hired out for a few pounds a day. With a bending machine it is possible to form all kinds of weird and wonderful shapes. A few simple operations are explained here, but if you struggle to pull a 90° bend to the accuracy described, do not worry, just pull the bend with a little extra spare pipe and cut it to the required length as necessary.

Forming a bend

The procedure described here can be used to form a bend at any angle up to 90°.

1 First, measure and mark onto a straight length of pipe the distance to the back of the bend you require, as shown in figure 6.6.
2 Place the pipe into the bending machine with this mark square in line with the back of the bending machine.
3 Attach the hook of the tube stop onto the pipe.
4 Position the back guide on the pipe and engage the roller to hold it in place.
5 Finally, pull down the lever arm to form the bend, stopping when the desired angle is achieved. Note that when forming bends in 22 mm pipe, considerable strength is required.

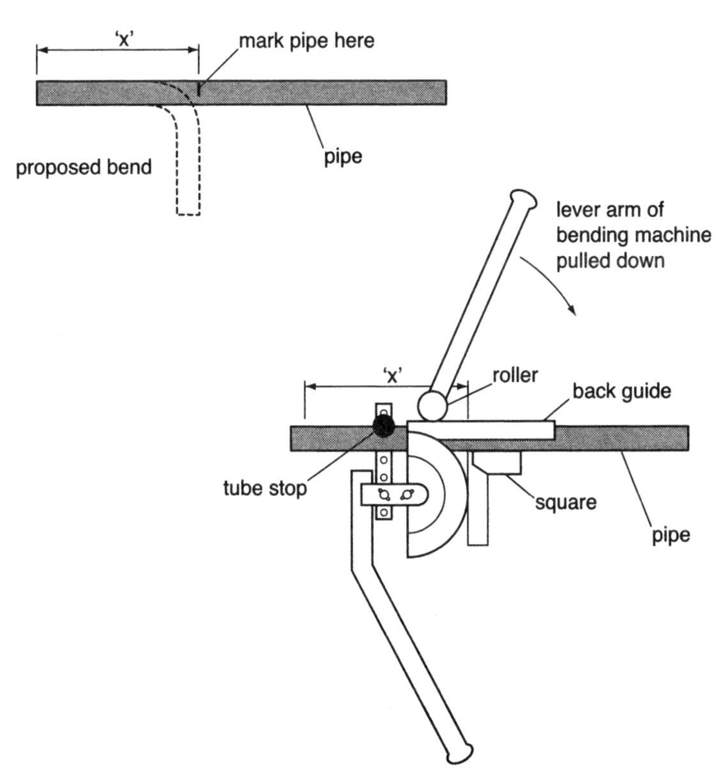

figure 6.6 pulling a 90° bend

Forming an offset

An offset is a series of two bends that, in effect, allows the pipe to continue in the same direction but along a new parallel plane. This is achieved as follows:

1 Take a measurement of the required offset.
2 Now pull the first bend to an angle within the machine. This angle can be as large or as small as suits your needs, but should not be too sharp otherwise there will be insufficient room for the tube stop and hook to sit onto the pipe when making the second bend. An angle of around 30° is usually about right.
3 The pipe is now repositioned in the bending machine, with the bend just pulled pointing upwards. Ensure the pipe is lying in the bender with the first offset in line with the direction of the roller, otherwise your second bend will be pulled on a different plane. A straight edge is now used, placed parallel to the angle of the first bend formed, to measure the required distance of the offset (see figure 6.7).
4 Once you have measured the correct distance for the required offset and put the tube stop in place, the pipe can be pulled round until the correct angle is achieved along the new parallel plane.

Plastic pipework and fittings for hot and cold pipework

Over the last 15 years or so there has been quite a revolution within the plumbing industry regarding whether to use copper or plastic for the pipework within the building. Plastic pipework today can be used safely for both cold and hot water supplies, including the central heating. Plumbing systems can certainly be installed much more quickly with plastic piping, and no jointing is required in long pipe runs. It is also easier to poke or push it through difficult locations or pipe ducts. Water noises due to water flowing through the pipes are also greatly reduced. But for pipes that will be seen and run on the surface, plastic looks rather messy. It lacks the sharpness and conformity of a regular shape that one expects from a piece of copper tube.

Fortunately, nowadays the external pipe diameter of most plastic pipe is the same as for copper and therefore you can simply use a mix of the two materials, running plastic below floors and anywhere else they won't be seen, and making the

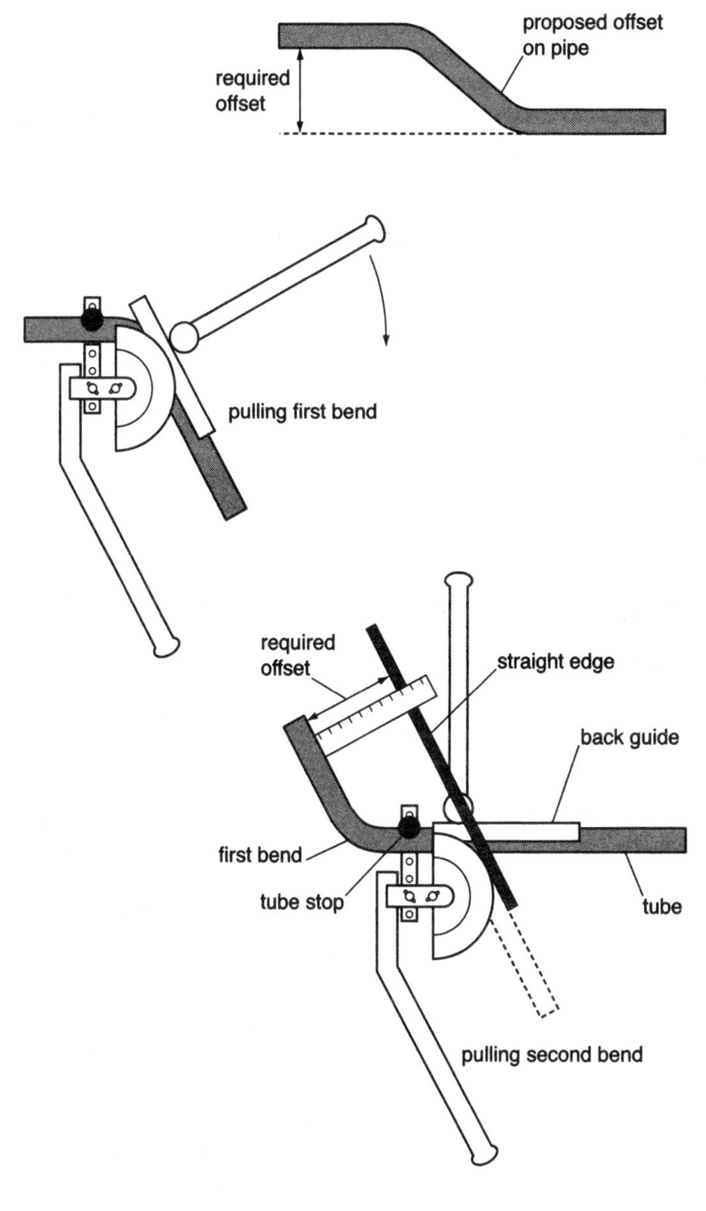

figure 6.7 forming an offset

final connections that will be on show in copper. The push-fit method of jointing would be used (see above).

The polyethylene (PE) plastic pipe used underground, such as that used for the mains water supply pipe from the road into the building, is of a different type and to make this type of joint a special compression fitting is usually made, although there are push-fit joints that can be used. It should be noted that, when making this plastic joint, an internal sleeve is inserted into the tube end as it enters the fitting, thereby providing additional support. Polyethylene pipe has a very thick wall so, for example:

- 25 mm PE equates to 22 mm copper pipe
- 20 mm PE equates to 15 mm copper pipe.

Plastic waste pipework

Within the domestic home, plastic plumbing materials for internal drainage pipes have been used now for well over 40 years. These materials are very simple to join together and when installed correctly last for many years without any problems whatsoever. Three types of joint are used:

- push-fit
- solvent weld
- compression fitting.

Push-fit joints

These consist of a large 'O' ring housed within the fitting and into which the spigot of another fitting or plain end of a pipe is pushed. In order to make a successful joint the pipe needs to be cut square and a small bevelled edge is chamfered onto the pipe end, using a rasp or similar tool. Now, ideally, some silicone lubricant or soap solution is put onto the pipe and it is pushed firmly into the fitting. Where a long pipe run has been made it is advisable to re-pull the pipe from the fitting a little, say 10 mm, thereby allowing for expansion and contraction of the plastic pipe itself.

Solvent weld joints

These joints, once formed, cannot be reused, unlike the push-fit joint which can be pulled apart and used over and over again. The solvent-welded joint uses special solvent weld cement. It is not a glue used to stick the two surfaces together but a solvent that burns into the pipe and fitting thereby bonding the two to form a sound firm joint. Once made, the joint hardens within seconds and when fully set no amount of pulling or twisting will have any effect. The joint is easily formed as follows:

1 First, clean the pipe end and the internal surface of the fitting with a solvent cleaner. This process can be omitted with reasonably clean fittings and pipe.
2 Now smear a thin layer of solvent cement onto the pipe end and inside the fitting to be joined to it. Bring the two together quickly, giving a slight twist, thereby ensuring the cement is in contact with all parts of the mating surfaces. Before the solvent sets, make sure the bend, if used, is facing in the desired direction. The fitting is then left to stand for a few minutes, by which time it will be set quite firm and will generally be ready for use.
3 It is essential that not too much solvent cement is used as the excess cement is pushed into the pipe and wasted, plus it will take the joint much longer to set. The solvent cement used gives off vapours so should not be used in confined spaces without plenty of ventilation, and the cement itself is also highly flammable.

Compression joints

Waste pipe compression joints are generally restricted to the connections of traps to the pipework. For this joint a rubber compression ring is used. The joint is formed as follows:

- First, push the nut onto the pipe.
- Then push on the flat plastic washer.
- Then push on the rubber compression ring.
- Next, fully insert the end of the pipe into the fitting, making sure it reaches the stop.
- Push the compression ring along the pipe to the mouth of the fitting.
- Now wind the nut onto the thread of the fitting in a clockwise direction. This pulls the flat washer up to the compression ring, forcing it into the fitting. These joints are generally made watertight by no more force than that required to tighten the nut up by hand.

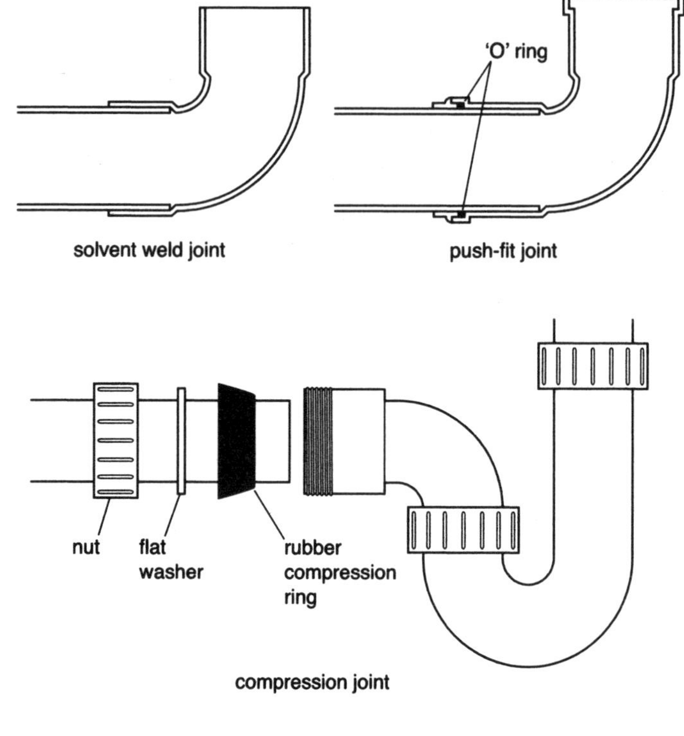

figure 6.8 joints used on plastic waste pipe

Specialist plumbing tools

Plumbers do have quite a full selection of different tools at their disposal, ranging from screwdrivers, hammers to blowlamps, yet most of these are not specialist as such. A plumber would have several spanners, usually of the adjustable type, and a pipe wrench or two. The pipe wrench is similar to the adjustable spanner except that the jaws have teeth, designed to grip pipes etc. Of course the bending machine is a specialist item, and when replacing an immersion heater a specialist extra-large immersion heater spanner is required, but the two most common 'specialist' items that the domestic plumber uses on a daily basis are the basin spanner and tube cutter.

The basin spanner

This tool is absolutely essential if you need to tighten or loosen, as the case may be, the nuts located up behind a bath, basin or sink where space is very restricted. There are several designs of basin spanner and the design selected is very much a matter of choice, although I personally find the adjustable wrench type the most versatile. Explaining how to use this spanner is very difficult and it is really necessary to get some hands-on practice. The turning direction, i.e. clockwise or anticlockwise, of the adjustable wrench shown, can be changed simply by altering the direction to which the toothed head is facing at the top of the shaft.

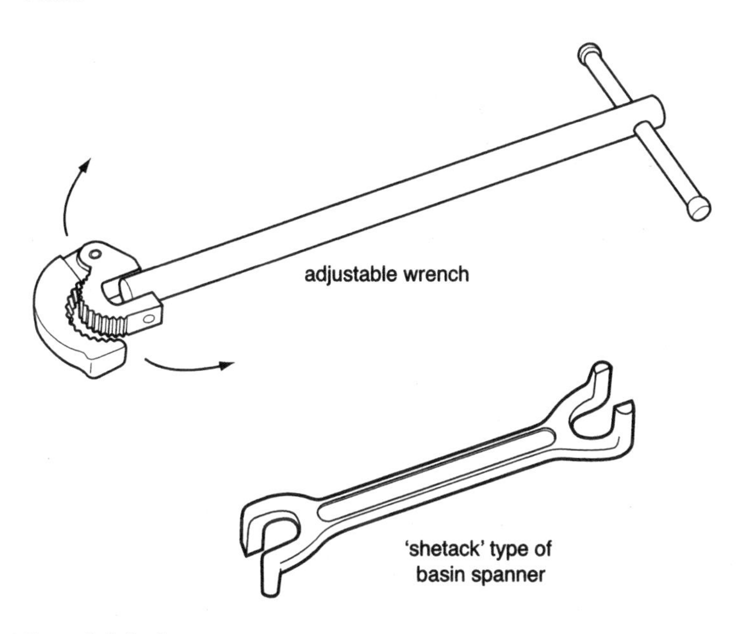

figure 6.9 basin spanners

The tube cutter

This is not an essential tool, as cutting a pipe can also be achieved with a hacksaw, in particular a junior hacksaw. But these are tools that will cut the pipe squarely and with a great deal of ease. Their biggest drawback is that they put a small internal burr on the pipe. Often the plumber does not worry about this, but it can cause noise problems which are not identified until it is too late to do anything about it. The internal burr should ideally be reamed or filed out and many cutters include a reamer for this purpose. The cutter is operated by winding down the handle until the single roller touches the pipe. The tool is then rotated fully around the pipe; the handle is then wound down another $^1/_2$–1 turn and rotated again. This process is repeated as many times as necessary until sufficient depth has been cut into the pipe to cause it to part. A particularly good cutter for getting into tight areas is the pipe slice, but with this you need to select one of the correct size, i.e. 15 mm or 22 mm. This design of cutter automatically cuts the pipe as it is rotated and no adjustment of the blade depth is required.

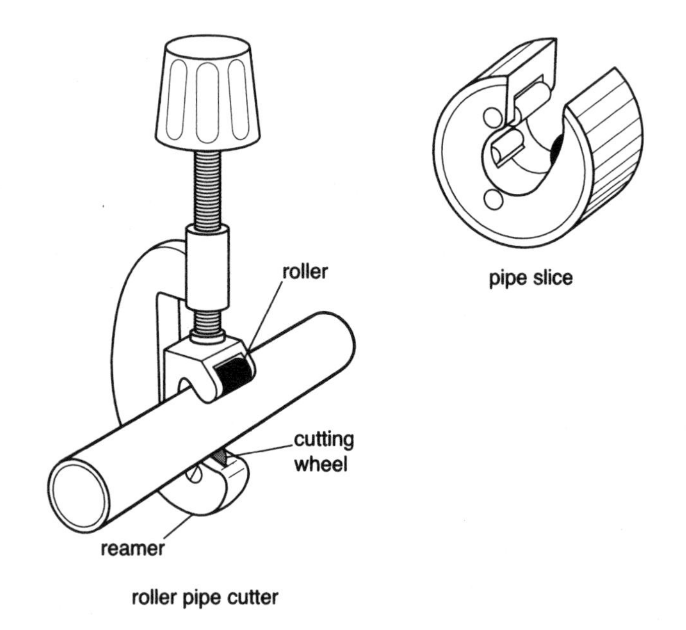

figure 6.10 copper tube cutters

These cutters will cut right through the pipe, so it is absolutely essential that you check that there is no water within the pipe before you cut it, otherwise this will flow uncontrollably from the pipe ends when they part.

Temporary continuity bonding wire

Although pipework these days is supposed to be bonded and safe from electrical currents (see equipotential earth bonding in Chapter 01), it is possible that there might be a fault, unbeknown to anyone, in which an electrical current is flowing down to earth through your metal pipework. Anyone who cuts the pipe and pulls the two sections apart runs the risk of being electrocuted. Plumbers rarely use a temporary bonding wire and are even more rarely electrocuted, but it does happen (on only half a dozen or so occasions a year), sometimes fatally. The choice is yours. What happens is that the fault current flowing down through the pipe to earth is interrupted as the pipe is cut. As the operative holds onto the two separate pipe ends the current can resume its path and flow through the individual, up their arm, trough the trunk and heart and back down the other arm to rejoin the pipe. Their muscles will contort with the shock and they will grip the pipe tightly and be unable to let go.

In order to ensure complete safety, anyone doing this sort of plumbing work should place a temporary bonding wire across the section to be cut, so that in the event of a current flowing the fault path is maintained as the two pipe sections are pulled apart. This bonding wire should be kept in place until the pipe section is reinstated, such as when inserting of a new tee connection. A bonding wire is essentially the same as a set of car jump start leads (see figure 6.11).

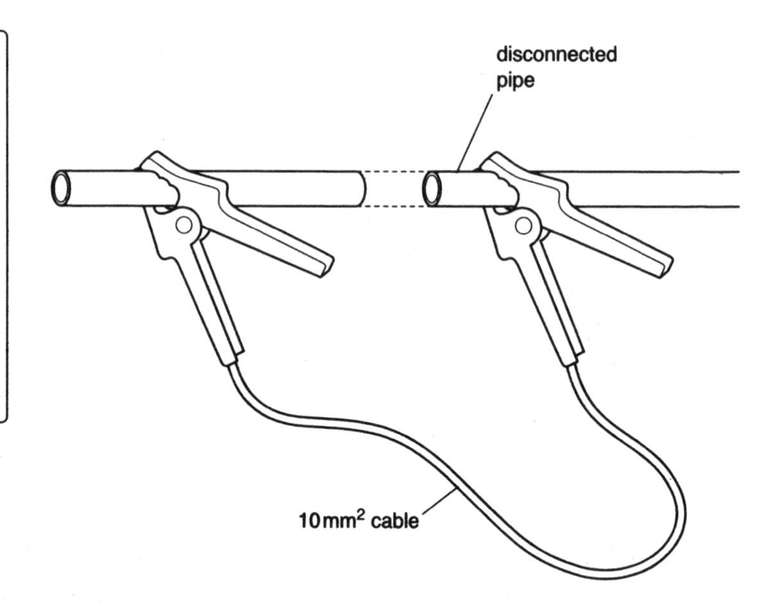

figure 6.11 temporary continuity bond

Concealing your pipework

Most people do no like to see pipework, so concealment is one of the keys to a successful plumbing job. Pipes run on the surface can never be made to look good, so hiding them within walls and below floors should always be considered. However, there are a few specific requirements that need to be observed. Figure 6.12 shows methods that can be used.

Pipes run below floors

Solid floors

There is no problem in running the pipework within the floor screed (i.e. the top layer of sand and cement) providing there is some protection around the pipe to prevent chemical attack or corrosion caused by the cement. In the case of heating pipework there also needs to be some provision to allow for expansion. This can be achieved by placing the pipe within some thin lagging material or running it within a small floor duct, covered with a plate. If you wish to run the pipe in concrete it will need to be fully protected and to do this you could run it within a larger sized pipe.

Timber floors

It is essential to remember that if you cut too much material from a structural floor joist you will weaken it, possibly making it unsafe. For example, the maximum depth to which a floor joist can be cut is one-eighth of the overall depth of the joist, and the notch should be made close to the bearing wall. Also, when running pipes below timber floors remember to allow for expansion and contraction and possibly consider laying the pipes onto felt pads to cut down the noise from these movements. You should also avoid pipes touching each other as this will also create noise problems.

Pipes within walls

Solid walls

Pipes can be concealed within an internal wall within a pipe chase (a channel cut into the wall, as seen in figure 6.12) and simply plastered over; however, there must be provision to isolate the pipe should a leak occur. Again, ideally the pipe should be protected against acid attack from cement-based products. As with floor joists there is a maximum depth at which any pipe chase can be placed before weakening occurs – this depth is one-third of the thickness of the wall for vertically installed pipes, and one-sixth where the pipe chase is run horizontally.

Timber walls

When running pipework within timber stud walls you must consider the possibility that the water flowing through the pipes could resonate through the structure. Securing the pipe clips onto rubber or felt mountings and adding additional pipe insulation material will help to reduce this. Above all, ensure that the system is fully checked for leaks before finally sealing in the pipes.

In all cases, wherever pipes will be inaccessible once they have been installed, joints should be kept to a minimum as these are generally the weakest point of the system and are the most likely to cause problems. Where possible, fit an access panel screwed in place to enable future access if required.

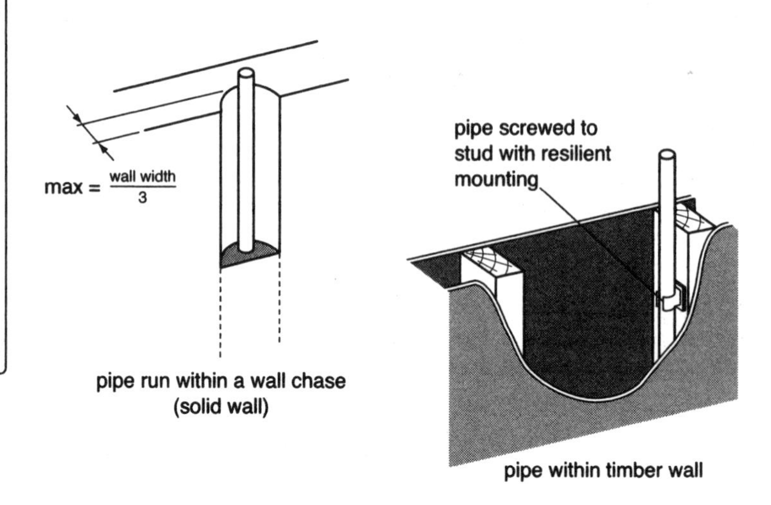

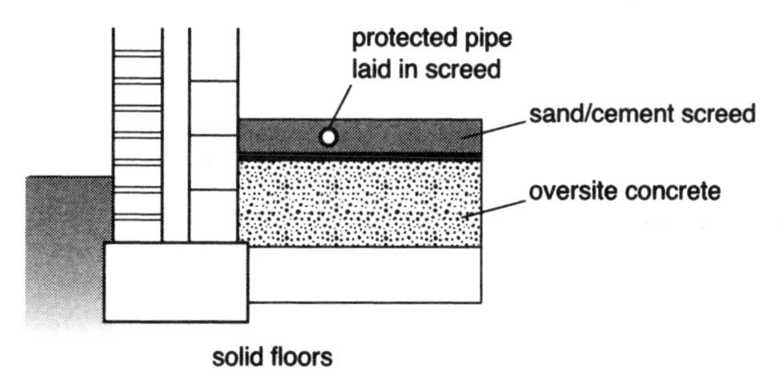

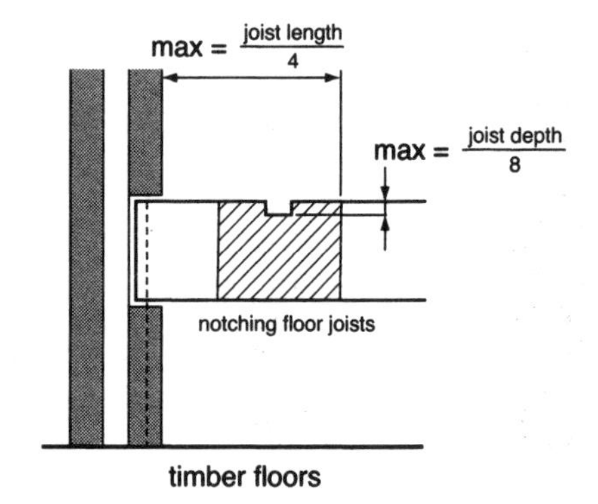

figure 6.12 concealment of pipework within walls and floors

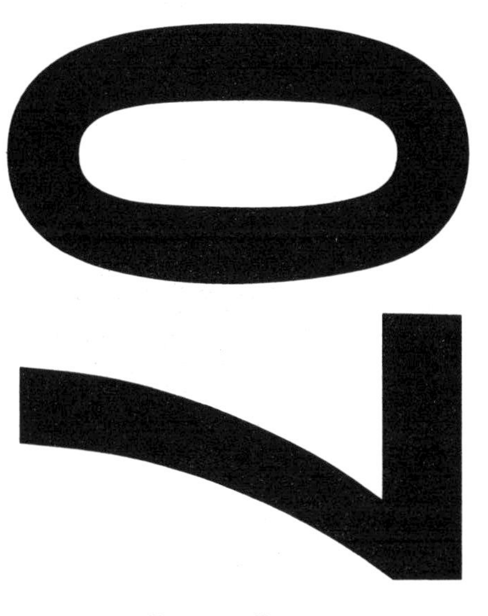

07 ancillary works and maintenance

In this chapter you will learn:

- about employing a qualified professional
- about gas or oil boiler maintenance
- about gas fire maintenance
- about general plumbing maintenance
- about maintenance of unvented hot water systems.

This chapter aims to look at some of the service and maintenance works that should be undertaken in order to keep your systems free from too many problems. It will provide an insight into what should be done by a professional, if they are called upon, and identifies what you should be looking for when employing someone to do this work for you.

Employing a qualified professional

Plumbing work around the home is generally within the grasp of anyone. With manufacturers making the assembly of components easier and the costs of employing a professional to do the work increasing, it seems sensible to look for other options such as DIY. However, there will come a time when you wish to call someone in to do the work. So, who do you call?

Unfortunately, many people think that because they can join two pipes together or can wire up an electrical component, they can now earn a living and trade as a 'professional'. However, a professional plumber or electrician has undergone extensive training in system design and has been tested on their ability to do the work.

Running pipes and cables of an incorrect size or following the wrong route can lead to problems that will probably be long term and recurrent, and may even be dangerous. Untrained operatives will undertake work that fails to perform as it should and that, due to ignorance or, dare I say it, deliberate breach of the law, infringes the various regulations in force.

There is no law to stop anyone trading as a plumber or as an electrician, but there are laws in place that require most work activities to be certified as completed correctly and to the right standard.

Certification

Over the last decade or so there has been a shift in the law, putting more and more onus upon the house holder to take responsibility for what they have in their home. If it is your home then *you* may be liable to prosecution for works completed that fail to comply with current regulations. This could also prove problematic when selling your home – as this legislation becomes more established and Home Information Packs (HIPs) become the norm, certification for work completed is likely to become a requirement.

A professional will usually be registered with a national validation body, such as one of those listed in **appendix 3: taking it further**, allowing the work to be undertaken immediately and permitting self-certification. Anyone else they may be legally bound to seek approval 'in writing' to do the work. If you do not do this you risk breaking the law with your installation. **Appendix 1: legislation** highlights the legislation that currently applies, and failing to employ the right person might result in you having to try and get work certificated at a later date, or failing to be in compliance with the small print of your home insurance policy should you wish to make a claim.

Professionals have to ensure that at all times they are in compliance with changes in legislation, and to achieve this they invariably need to:

- pay annual fees to belong to a professional body
- pay for continued training and assessment
- take time off from work and therefore lose earnings to attend courses
- comply with additional safety laws, which have additional cost implications.

All this has had a knock-on effect and caused an increase in what you would expect to pay for an hour's or a day's work.

Gas installation work within your home

It is illegal for anyone to undertake gas installation work for monetary gain unless they are registered with CORGI (see **appendix 1: legislation**). Employing someone who you know is not CORGI registered could also be deemed an offence. It is essential, therefore, that you always check to see the operative's current CORGI registration card, identifying on the back of the card what gas work they have been assessed to undertake.

Electrical installation work within your home

Again, all electrical work undertaken must be certified as safe – without this certification you cannot be absolutely sure that the work has been carried out by a competent operative. For example, a central heating installation has an electrical supply and therefore its installation legally requires a minor works certificate as a minimum.

So, where do you go to get the right plumber, heating engineer, gas engineer or electrician?

- One of the best options open to you is through a recommendation.
- Next, try contacting one of the recognized professional bodies such as those listed in **appendix 3: taking it further**. These bodies generally maintain lists of trading operatives local to the area where you live or where you require the work to be done.
- Do not necessarily go for the big ads in trade indexes such as the yellow pages. These may or may not be any good. My view is that a good company does not necessarily have to advertize for work.
- Don't accept the first quote you are given for a job; try to get at least three estimates where possible.
- Don't necessarily go for the cheapest option and don't have the work completed by someone who cannot offer the full services and, where necessary, the essential certification as identified above and in **appendix 1: legislation**.

Gas or oil boiler maintenance

Boiler manufacturers usually recommend that boiler servicing is undertaken on an annual basis, so I could not possibly recommend anything less. If you don't touch the boiler it may work for years, but will it be working safely and efficiently, that's the question!

Servicing is not carried out just to ensure that the boiler stays working; it is also to ensure that it stays working safely. Without an annual check-up, the odourless combustion products from a faulty appliance may discharge into the room.

Combustion products do accumulate on the heat exchanger within the appliance and in so doing reduce its efficiency. Most modern gas boilers have very compact heat exchanger fins which can rapidly block up.

There are all kinds of service contracts that are provided by installers at a complete range of costs. Some companies seem to do no more than stick a flue gas analyser into the flue to take an efficiency reading of the appliance and measure for carbon monoxide (CO). They might only consider undertaking a full service if the reading is too high, but this is not a service and is sometimes called a safety check. However, the sample flue gas products obtained will only be accurate at the time of taking the reading. If there is plenty of fresh air around this may lead to a good reading; conversely, if no fresh air can get into the appliance the reading may well deteriorate and worsen over time.

A service is a full check of the running condition of the appliance as recommended by the manufacturer. Ideally it would also include undertaking a flue gas analysis before and after the service. The service itself would include, among other things:

- a check on the gas burner pressure or oil pressure, as applicable
- the correct ventilation or air supply to the appliance
- inspection and, if necessary, cleaning of the heat exchanger
- inspection and, if necessary, cleaning of the burner head
- an inspection to confirm the correct fluing arrangements
- cleaning out, where applicable, the condensation trap as found in a high efficiency or condensing boiler
- checking the flame ignition and flame failure devices
- checking pressures or levels of expansion vessels
- correct operation of thermostats
- checking pressure sensing devices
- checking the system controls for correct operation
- and, above all, checking the safe operation of the appliance.

There is no law to state that you cannot service your own appliance in your own home. But would you be sure that you have an appliance that is operating safely? Oil engineers registered with OFTEC and gas service engineers registered with CORGI have been assessed on their ability to undertake work on these appliances. Ask what their service consists of, and whether they will be giving you a service or safety check. There is no law to say a certificate should be given for a service, but you should try to find out what they will be doing for their money.

The boiler is the item everyone thinks of when considering appliance servicing within the home, but there are other appliances and systems that are also worthy of consideration which we look at now.

Gas fire maintenance

Gas fires are often installed and run for years without anyone checking them. Some of the old gas fires still in use today have been in operation for well over 20 years, possibly never having been inspected since the day they were installed. Over this period they will have suffered the strain of time and will invariably have cracks in the heat exchanger, unseen by the user, and products of combustion may be drawn into the room. In Chapter 02 the effects of carbon monoxide (CO), including the possibility of death, are discussed.

Also consider the poor little birds outside in the winter! Where would you sit if you were one? On top of the chimney pot is a nice warm spot, and it also makes a nice place for them to leave their droppings. Unfortunately birds do not live for ever and inevitably they sometimes fall down the chimney. All the droppings, dead birds and other material such as leaves accumulates within the chimney and eventually creates a blockage within the flue system from the fire.

The gas fire installed to the flue will still operate, but will discharge its products into the room rather than up the flue. Carbon monoxide has no smell so you won't even know that you are being poisoned. If you look at table 2.1, relating to carbon monoxide poisoning, you will see that there only needs to be as little as less than 1 per cent within the room before a fatality will result within a few minutes.

It is important occasionally to have your gas fire checked out by a professional; it may save you life! An expert will not only check that it is operating safely but will also check the operation of the flue. Expect the service engineer as a minimum to:

- Remove the fire for inspection and, if necessary, clean the burner head.
- Check the condition of the radiants and heat exchanger, replacing any faulty parts.
- Check for the correct ventilation of the appliance.
- Check for the correct fluing arrangements, removing any debris.
- Undertake a flue flow test, which consists of passing a quantity of smoke up through the chimney to confirm that it is clear.
- Reinstate the appliance, ensuring a secure fixing and that the flue seals are maintained.
- Check the gas burner pressure.
- Check the flame ignition and flame failure devices.
- Undertake a spillage test while the appliance is in operation.

The spillage test mentioned above is one of the most important tests and is undertaken by holding a special smoke-producing match at a position around the canopy just above the flames or radiants. If the products of combustion spill into the room then the smoke will likewise be pushed back into the room; conversely, if the products are being drawn into the flue the smoke likewise will be sucked into the chimney.

One of the key indicators of a gas fire continually spilling products of combustion is black staining to the walls or on the canopy of the fire, just above the flame itself.

Decorative fuel-effect gas fires

These fires have been around now for some 20 to 30 years. In the early days they were put together quite precariously, often consisting of a bent copper tube with a series of small holes drilled into it and laid in a bed of dry sand, through which the gas could filter and escape, creating the effect of naturally burning coal or wood.

The design of these fires has improved considerably since then and fires of this nature fitted these days even have their coals laid out onto the fire bed in a systematic order. I mention these fires because often the user has not been made aware that they are 'fuel-effect' fires, not real solid fuel fires, and as such you should not throw items such as paper, cigarette ends and so on onto them to burn. Yes, of course these items will burn, but in so doing they pass combustion products, including carbon or soot, into the flue system where they accumulate, altering its effectiveness. Also, these items will leave deposits of ash, which will not be cleared away as they would with real coal/wood burning fires. This ash accumulates and disrupts the correct operation of the fire, again possibly leading to the products escaping into the room.

There may be other gas appliances within the home, such as a water heater or cooker. These, although rarely the cause of any fatalities, should still occasionally be given a safety check.

General plumbing maintenance

Plumbing in the loft

This is an area of maintenance that is often overlooked. Unless the cistern overflows no one generally goes near the cisterns.

However, modern overflow pipes do have a filter contained within the housing that connects to the cistern wall. The lid will also have a filter contained within its vent.

It is quite unlikely that these filters will become blocked with insects; however, the time to find out is not when the float-operated valve does not close off properly, requiring the cistern to the overflow. Therefore, occasionally inspect the overflow and lid filters:

- Hold the ball float down below the water level to allow the cistern to overflow for a moment. This will confirm the filter is free and that the overflow is effective, without any leakage.
- Check the security of the lid.
- Confirm that the insulation material is held securely in place.
- Check that the isolation valves are operating freely.
- Look for general signs of fatigue or damage, particularly with older galvanized cisterns.

Cold and hot water supply stopcocks and valves

All the internal valves used for turning off the hot or cold water pipework should occasionally be operated. This prevents them from seizing up and keeps them operational in the event of an emergency. Check also that the label identifying what services it supplies is still in place and legible.

Maintenance of an unvented system of hot water supply

These systems do have the potential to blow up!

- If the pressure and temperature relief valves fail to operate along with the temperature thermostats, the pressure will continue to rise within until it can take no more.
- The pressure in these systems has the potential to increase above 1 bar (this being the pressure created by the atmosphere surrounding the system). Water boils at 100°C at 1 bar pressure; at greater pressures the boiling temperature of the water increases.
- Because of this, if a fracture were to occur in the cylinder when the pressure is above 1 bar, all the water would instantaneously flash to steam as it came under the influence of atmospheric pressure.

- When water changes to steam it expands around 1600 times, so the damage from such a steam explosion would be catastrophic.

There are several fail-safe devices incorporated within this system. As a minimum the test levers should be checked to ensure that they are still functioning.

In an ideal world these should be checked annually but in reality this task, like many maintenance tasks, is wrongly put off until another day. I would not recommend putting off the maintenance of these controls indefinitely.

Unfortunately, the biggest problem is that when these test levers are operated they invariably let-by (in other words the valve does not close off properly and it drips) and continue to drip. The reason for this may be that there is some limescale building up beneath the valve head, which suggests the valve needs replacing in any case. Changing the valve should be undertaken by a qualified operative as technically these are the only people, identified under the Building Regulations, with the proven competence to replace them.

Should a specialist be called in to check your system as part of a maintenance contract, in addition to checking that these test levers operate they should as a minimum also check:

- the pressure/volume capacity of the expansion vessel
- the in-line filter for debris
- that the normal operating thermostat is functioning, closing off at 60°C maximum.

Typical problems generally encountered with these systems include the following:

- Water discharging intermittently from the pressure relief valve. This is generally due to a pressure build-up within the system, possibly caused by:
  - a faulty pressure reducing valve
  - a faulty expansion vessel (this will most probably have lost its air pressure charge), thus as the water is heating up it cannot expand into the expansion vessel and forces open the pressure-relief valve. Where the air charge pressure is lost from a sealed expansion vessel the system will need to be drained down and the vessel re-pressurized.

- Water discharging continually from the pressure-relief valve. This could possibly be caused by:
  - a faulty pressure-reducing valve
  - a piece if grit lodged beneath the outlet valve seating of this control.
- Water discharging from the temperature-relief valve. This could be the result of:
  - a piece if grit lodged beneath the outlet valve seating of this control
  - both the normal operating and high-limit thermostats failing to operate.

In all of the cases above it is essential that the cause of the problem is investigated and that the water seepage is not simply plugged off with a fitting to stem the water flow. It is telling you something may be wrong!

undertaking small plumbing projects

In this chapter you will learn:

- **how to install a washing machine or dishwasher**
- **how to install a water softener**
- **how to install an outside tap**
- **how to remove a radiator to enable wallpapering behind it**
- **how to repair/replace the incoming cold water supply main**
- **how to install a new storage cistern**
- **how to repair a faulty immersion heater**
- **how to insulate to prevent freezing and frost damage**
- **how to install guttering and rainwater pipes**
- **how to install a range of sanitary appliances**
- **how to replace a shower booster pump.**

This chapter will provide guidance on undertaking some of the smaller plumbing tasks you may wish to tackle yourself. This section of the book assumes that you have read and understood the basic plumbing processes discussed in Chapter 06, and that you know how to turn off the water supply (see Chapter 04) and therefore have an adequate grounding to tackle these small projects.

You will soon discover, as your confidence grows, that by following the same basic principles and taking your time to think through how best to tackle a particular job that all kinds of tasks can be attempted. The instructions here allow you to sequence your activity and break down tasks into small, manageable chunks.

Invariably it is the preparation that takes most of the time. Taking up floorboards and making way for the pipe runs is the donkeywork – the running of pipes is often the easy bit. Start by getting rid of everything that's not wanted – it's going anyway and if you try and work round things it slows you down and prevents you running the pipework as you would choose to. With the old products out of the way you can easily gain access to areas such as those below floorboards or behind timber panels where the new pipework may be run (i.e. out of sight). Having cleared everything away you will have made room to work comfortably instead of falling over everything.

You will find that the same principles apply when installing a single WC suite as when installing a complete bathroom suite. So, in all of the projects to be undertaken the sequence of events will go something like this:

1 Turn off the water supply and hot water heaters where applicable.
2 Confirm that the water has shut off correctly.
3 Remove items that are no longer wanted.
4 Temporarily cap off the pipework at a point where the new connection is to be made, thus allowing you to turn the water supply back on until the new work is ready.
5 Prepare the work area for the new installation, running all new pipework and installing the appliances.
6 Turn off the water supply and make the new connection as necessary.
7 Turn the water back on and test the new appliance as appropriate.

Prior to undertaking any of the following tasks I would recommend that you read this chapter in its entirety as you will discover that all of the jobs have similarities and you may find that tips contained within one entry are also applicable to another.

If you require some expert advice or the services of a professional, see **appendix 3: taking it further**.

Installing a washing machine or dishwasher

This is one of the simplest projects to undertake as an introduction to doing your own plumbing works. The requirements for a dishwasher are the same as those for a washing machine so, in effect, these notes are applicable to both appliances. When you purchase the machine it will come with its own set of installation guidance notes that can be used to support what is written here. You will need the following to complete this task:

- adequate space into which the appliance can be fitted
- an electrical supply point within reach of the appliance cable
- a drain within close proximity or along the same stretch of wall
- a hot and cold water supply, the closer the better*.

Assuming the first two points are fulfilled, all that needs to be done is to run the waste pipe and water supply connection.

The waste pipe

If you are lucky you are installing the appliance next to an existing sink. Look at the trap from this sink; if it is plastic it may already have a special washing machine trap fitted for a branch connection to a washing machine or dishwasher (see figure 8.1). If this is the case, all you need to do is cut the blanked end off to enable you to push the washing machine waste pipe hose onto the tapered connection and secure it in place with a large jubilee clip (available from plumber's merchants and hardware stores).

*Most machines will work with just a cold supply and some machines only require this, but if there is no hot supply the operating time of the machine will be longer as the water will have to be heated by the machine. Check with the machine supplier.

If there is not one of these traps there, the best option is to purchase one and replace the existing trap, altering the existing waste pipe to the sink if necessary. Sometimes this may require the replacement of the existing sink waste pipe, but this is often still the best option and overcomes the need for the additional waste pipe which would be required as the alternative. You should avoid cutting a tee branch connection into the existing waste pipe as is often done by the DIY plumber, as this may result in problems with trap siphonage (see Chapter 01).

Where it is not possible to make a connection to a washing machine trap a new waste pipe will need to be run. To do this you would need to gain access to a drain connection. Look outside the building for a gully into which you could discharge your new waste pipe.

Alternatively, you may need to make a boss connection into the soil stack as it passes down to ground level. This is a much bigger job and for this you will need to see the notes on making a branch connection to the existing vertical soil pipe (page 194).

If you have decided to run a completely new additional waste system for the washing machine, the pipework must be run following the guidelines for waste pipework as discussed in Chapter 01, with a minimum waste pipe size of 40 mm (1½"). The new waste system will need a trap to be fitted and the pipe terminates with an upstand, as shown in figure 8.1. The washing machine waste pipe is then simply hooked into this upstand.

The water supply connection

This is made to the flexible hot and cold hose/s from the appliance. To make this connection you need to terminate your pipe with a quarter-turn washing machine valve, one for the hot water and one for the cold, within about 300 mm of the back of the machine. As with the trap, look first to see if these are already there. Where these are not present a new connection will need to be run.

To complete this task you need to:

1 Locate a suitable pipe where you could cut in your new tee joint.
2 Ensure you have chosen the correct pipe for your connection by following the route of the pipe to check that it feeds a hot or cold outlet as necessary.

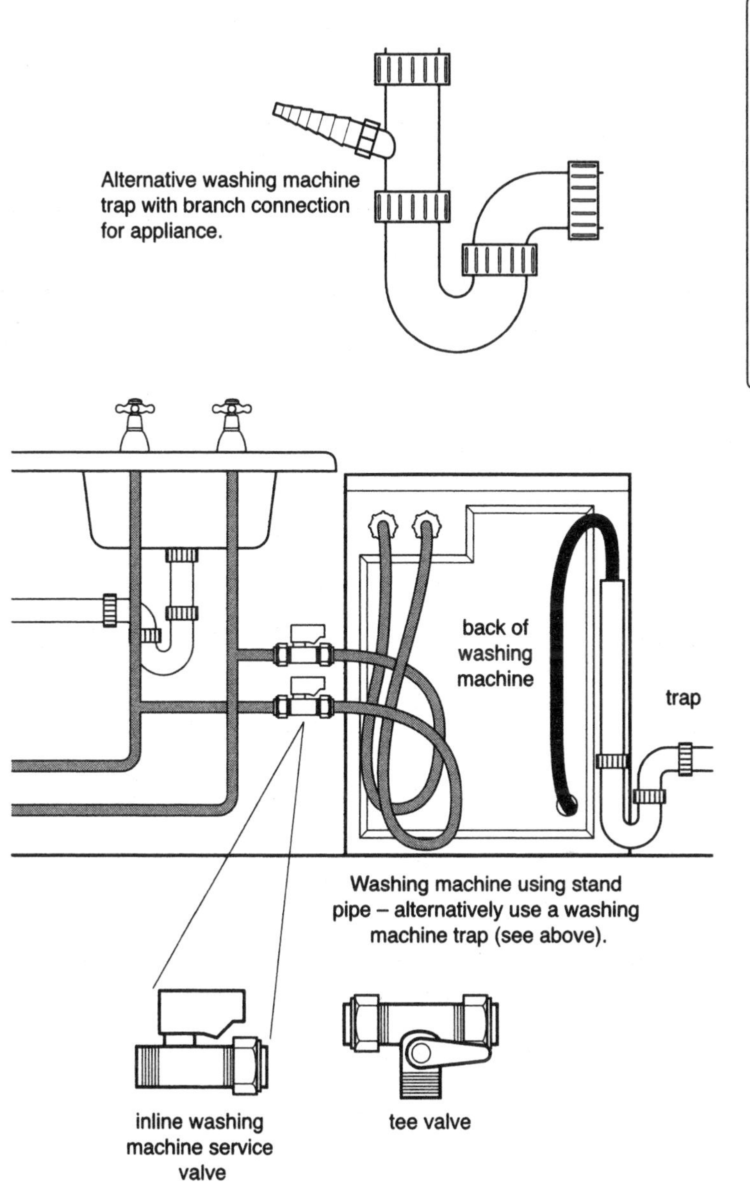
Alternative washing machine trap with branch connection for appliance.
back of washing machine
trap
Washing machine using stand pipe – alternatively use a washing machine trap (see above).
inline washing machine service valve
tee valve

figure 8.1 installing a washing machine or dishwasher

3 Install your new pipework, terminating with the quarter-turn washing machine valve close to the washing machine. These can be purchased with a red or blue head as for the hot and cold supply. If there is a pipe close to the machine into which you could cut, and there is sufficient room within the pipework, you can purchase a special tee valve for this purpose.
4 Note that the new pipework is run prior to making the final connection to the existing supply, thereby ensuring that the water is only turned off for the minimum amount of time.
5 Run the new pipework following the basic plumbing processes discussed in Chapter 06. Turn off the water supply and confirm that water is drained from the pipe by opening the taps along this section of pipe and/or a drain-off cock.
6 Cut into the existing pipe and make the final tee connection into the existing pipework.
7 Turn the water back on and turn on the pipe to test your work.
8 With the hot and cold supply valves in place you just need to make the final connection to the machine. This will be via two hoses, supplied with the machine, with a rubber washer making the connection at your new valves and onto the termination points on the machine. These joints should not be done up too tightly.
9 Turn on the quarter-turn valves and check for watertightness, doing up the nuts a little if necessary. Now, if you have not already done so, it is essential to remove the transit bracket that was secured at the factory to prevent the drum from moving and causing damage during the transportation of the machine.
10 Finally, plug the machine into the power supply and away you go.

Installing a water softener

The installation of a water softener is in itself a relatively simple task, but you do need to ensure that a hard water connection direct from the water main is maintained prior to the water softener connection. As with the installation of a washing machine, you will need:

- adequate space into which the appliance can be fitted
- an electrical supply point within reach of the appliance cable

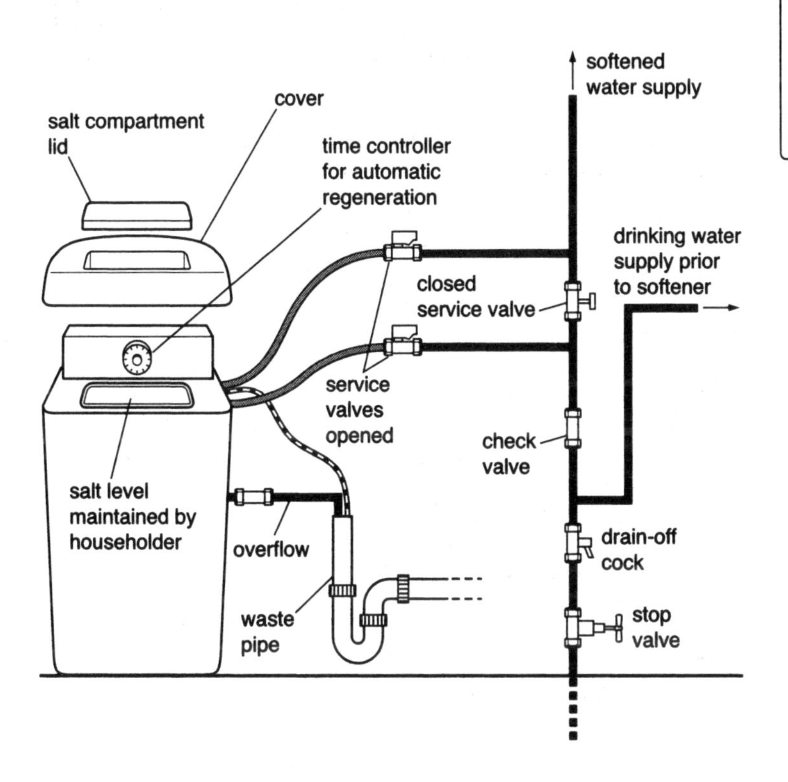

figure 8.2 installing a water softener

- a drain within close proximity or along the same stretch of wall
- a cold water supply, the closer the better.

When installing a water softener you will need to run the waste pipe. To do this, follow the guidance given above for running the waste pipe when installing a washing machine. The cold water connections are made into the water supply main as shown in figure 8.2 Note that a connection has been run to the sink to provide water for drinking purposes. Softened water is generally regarded as safe to drink, but it is not recommended for pregnant women and young children; some people also do not like its taste.

There will be installation instructions provided with the appliance, which you can follow. When cutting into the existing cold water mains supply pipe, as always, make sure the water is turned off and that any excess water has been drained from the pipe via the drain-off cock. Ensuring that the float-operated valve in the storage or toilet cistern is opened will help to drain down this pipe – it allows the air in to force the water out. The process of joining to the pipework is discussed in Chapter 06.

When the installation is complete you will need to fill the water softener with salt and set the time clock, following the instructions supplied by the manufacturer.

Making a branch connection to the existing vertical soil pipe

Sometimes, in order to make a connection to the drainage system, it is necessary to cut into the vertical soil stack. You will need to insert a tee branch connection for large pipes; however, where this connection is for a small pipe the connection is called a 'boss', as shown in figure 8.3. Ideally, when looking for a connection for a new waste pipe, try to find a gully or existing connection so that the amount of work is kept to a minimum (alas, life is rarely this simple).

The procedure identified here is for jointing into a plastic drainage pipe; making a boss connection to cast iron or asbestos soil stacks is beyond the scope of this book. Follow this simple process to make a boss connection to a plastic soil pipe:

1 The first thing to do is purchase a strap-on boss for the size of waste pipe you are using.
2 Once you have this you will see the size of hole that needs to be cut into the existing soil stack. This hole should be cut out with a special hole saw. This has a central drill bit and a circular toothed saw blade, purchased for a few pounds at any hardware store. Before cutting the pipe you check that the saw is the right size for the boss fitting. You will notice that a lug is designed to fit into the hole to be cut, thereby keeping it central. Cut the hole too small and the lug will not enter and will prevent full contact of the mating surfaces. Cut the hole too big and you run the risk of insufficient coverage of the mating surfaces.
3 With everything in place and access gained to the soil pipe, ensure the waste system is not used by anyone until you have finished your work. Then drill the hole in the correct location, following the guidance for slope gradient, as discussed in Chapter 01. You must not fit the boss too high or you will trap water in the pipe as it drains from the appliance.
4 With the hole made, confirm that the boss fits snugly to the pipe.
5 Now clean the mating surfaces, removing any paint or other residue from the existing soil stack to provide a clean plastic-to-plastic joint.
6 Apply the appropriate solvent cement, made and supplied by the manufacturer of the plastic strap-on boss.
7 Now push the fitting into place and hold it firmly until the cement has set, usually within a few minutes; some designs have a strap that passes right around the large soil pipe and bolts the strap onto the pipe.
8 Some designs of boss include a rubber 'O' ring to enable you now to push the small waste pipe into the fitting; others, such as the one illustrated below, now require you to insert a rubber cone prior to making your final connection.

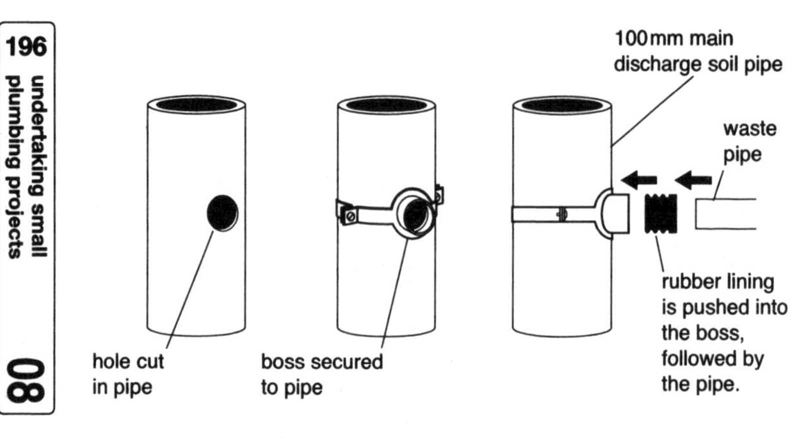

figure 8.3 strap-on boss

Installing an outside tap

The type of tap used outside is referred to as a hose union bib tap, having the provision for the connection of a hosepipe. Installing this tap is a relatively simply task – possibly the biggest job is drilling the hole through the external wall, through which to run the pipe. Installing an outside tap also requires some specific regulations to be complied with, as follows:

- There must be an inline stop valve fitted, such as a stopcock, to isolate the tap.
- A device must be installed within the pipeline to ensure that no water can be drawn back into the water supply pipeline should a negative pressure be created within. This device is called a double checkvalve, which basically incorporates two spring-loaded non-return valves.
- There must be provision to drain out the water in winter when the tap will not be in use.
- Thermal insulation material must be fitted as appropriate.
- Finally, it should be noted that when running the pipe to this external location there is the potential for stray electrical currents, due to a faulty electrical system, to pass to earth when someone outside touches the tap. Therefore, to avoid this it is recommended that a plastic fitting is included within the pipeline to the tap.

Installing the tap is completed as follows:

1 Drill the hole in the wall at the desired location.
2 Inside the building, run the pipe to the nearby cold water supply and install a stopcock or alternative isolation valve. Further along the pipe leading to the outside tap the double checkvalve and a drain-off cock are incorporated. Finally, assuming the work completed has been run in copper, before the pipe exits the property a plastic push coupling should be incorporated.
3 Run the pipe through the wall and terminate it securely, possibly to a back-plate elbow. This fitting is designed with a female thread into which the hose union tap can be secured and has the provision to be screwed back to the wall, which will hold the tap firmly in place.
4 With everything in place, turn off the water supply and confirm that it is closed. Cut the pipe to make a tee connection as necessary, branching to your new outside hose union tap.
5 Finally, once the pipework has been tested, insulation material must be applied to the pipework to provide adequate frost protection.

Removing a radiator to decorate behind it

Where a radiator needs to be removed from a wall in order to decorate behind it, this can be done quite simply without having to drain down the whole of the heating system. The job sometimes requires two people where the radiator is quite large.

The first task is to turn off the radiator at both ends. One end will have a lockshield valve attached, whereas the one at the other end will be manually operated or will be a thermostatic radiator valve (TRV).

Closing the lockshield valve

To turn this off, pull off the plastic cap/cover (sometimes a screw is located in the top to hold this cap on). With this removed you can now use a spanner to turn the spindle clockwise until the valve is fully closed. Take a note of the number of turns you make to close this valve – it may only be half a turn or it may take several turns. When reinstating the radiator it is important that you only open this valve the same number of turns as you used to close it, as this valve has been adjusted to balance the system, thereby ensuring water feeds equally to all of the radiators within the system. Opening it too much might affect the operation of the other radiators, in effect stealing all the hot water.

Closing the manually operated valve or TVR

You now need to turn off the valve at the other end of the radiator. This may be a manually operated valve, which is simply turned clockwise to close, or there may be a TRV fitted, in which case you will need to fit the manual isolation head that came with the valve. If you just turn down a thermostatic valve it may turn off the water, but if the temperature within the room drops it might automatically open again, allowing water to discharge onto the floor while the radiator is off the wall. There is a pin below the thermostatic control which needs to be held down, keeping the water from passing through the valve.

Removing the radiator

With the radiator isolated you must now confirm that the valves are holding back the water flow. Do this by opening the air-release valve at the top of the radiator with a small radiator key. Water will initially spurt out due to the pressure contained within, but should subside within about three to five seconds. If the water continues to flow you know one of the valves is not fully closed, so you need to check the two valves again. When you can open the air-release valve with no water flowing you will know that the valves are closed. Now it is *essential* to re-close the air-release valve, otherwise air will enter the radiator and force the water out onto the floor when you undo the union nuts at the base of the radiator.

Having confirmed that no water is flowing into the radiator you can now undo the large union nuts at both ends, which connect the radiator to the isolation valves. The radiator is still full of water at this point, but this can only come out if air is allowed in. However, you will need to be prepared for a little water to discharge, which may be inky black in colour. With both unions fully undone and the radiator disconnected from the valve you simply lift it from the radiator brackets and place your thumb over the open end, as quickly as possible, thus preventing air going in and the water escaping. You will need to be prepared to take the weight of the radiator and the water it contains, hence the need for a second person. Take the radiator outside and tip it up to remove the water. Now go back and check that the valves that were connected to the radiator are not dripping.

Reinstating the radiator

1 Apply a little jointing compound such as 'Boss white' between the two mating surfaces of the brass union, to ensure a sound seal when tightening back up the union nuts to the radiator. Do not use PTFE tape on the threads as these do not form the seal, but just act as the leverage to pull the two mating surfaces of the union together.
2 Turn down the room thermostat; this will turn off the power supply to the central heating pump, hence ensuring that air can be bled from the system without a possible negative pressure caused by the pump sucking air into the system.
3 Turn on the radiator valves and bleed the air from the top of the radiator with the special radiator key until water is seen to emerge. Remember, only open the lockshield valve the same number of turns as you used to close it.
4 Check the two union joints to confirm that they are not leaking.
5 Finally, turn the room thermostat back up to the desired temperature.

Repairing the incoming water supply main

A leaking underground water supply main is often left undiscovered for months. One of the key indicators is when you experience a lack of water flow or there is a continued sound of a water flowing from the mains pipework but you have no taps open. This is particularly heard at night when there is greater pressure within the mains and all is quiet around the house.

Where the leak is on the underground stopcock, such as through the packing gland nut (see Chapter 04), it is possible sometimes to reach down the pipe duct with tools such as the adjustable basin spanner (see figure 6.9) to cure the problem. But the leak may be at any point along the entire length of the pipe, so if it is not at the stopcock you will have to dig down to the pipe and expose it. You can only start by chancing a test hole where you think the leak might be. Once the pipe depth has been reached you often get a clue as to what direction to dig in due to the direction from which the issuing water is flowing. But make no mistake, there is no quick fix and you may have to search for some time.

Once you have found the leak, how you make the repair will depend upon the material used for the supply. In all cases you will have to turn off the supply from a valve further upstream, such as at your outside stopcock or by getting the supply turned off by the water authority.

If you have a polyethylene or copper supply pipe the repair is usually quite a simple process. It might entail remaking a joint that has come apart or it might be necessary to cut the pipe and insert a new section, joining the piece cut out with a new piece of plastic or copper pipe.

Where you have an existing lead or steel mains supply pipe, it may be more appropriate to consider replacing the entire length. It may be possible to insert a short section of plastic, using two specially designed compression couplings for this purpose. There are strict Water Supply Regulations preventing:

- the use of lead in new or repair works
- the use of copper upstream of lead or galvanized steel pipework, so this could not be inserted midway along a length of pipe run.

You need to remember:

- Where steel has been used it has already well exceeded its life expectancy.
- Where lead mains are found they should be replaced whenever possible on the grounds of safety due to the toxic nature of lead.

Replacing the supply main

If you plan to replace the entire length of the supply pipe, it may be worth considering the hire of a mini digger. The task is quite straightforward apart from the large amount of manual labour involved in digging out the hole and then refilling it. It is hoped that when you have exposed the entire length of the supply main you will discover that the builder included a pipe duct in the foundations of the building through which to pass your pipe into the house. For buildings over 35 to 40 years old, however, don't bank on the pipe duct being there and you may need to undertake additional laborious work cutting through the foundation wall and down through the floor in order to make a route for your new pipe.

With the route exposed (minimum 750 mm depth – see figure 1.1) between the outside and internal stopcocks the new recommended 25 mm polyethylene pipe can be laid within the trench. The pipe should be run in one complete length, avoiding coupling or connection joints. It is a good idea to lay the pipe within the trench from side to side, thereby providing some spare pipe to allow for expansion and ground movement. For the pipe connections that need to be made at each end of the pipe run, see the section that deals with pipe jointing in Chapter 06.

Installing a new storage cistern

Storage cisterns used today are made of plastic and if you are going to install one it is essential to ensure that the base is completely supported, otherwise the weight of water contained within the cistern will cause the plastic to stretch and eventually break at the unsupported point. Old galvanized cisterns did not require this total support.

These old metal cisterns are invariably left in the roof space as removing them requires extensive additional work. Sometimes, old asbestos cisterns are encountered. These are fine while in use, but when they have passed their useful lifespan it is essential that they are disposed of safely. *Do not* cut the material, as this will make a dust which is extremely dangerous to inhale; even a minute particle can be hazardous to your health.

The size of a new cistern should be a minimum capacity of 100 litres if it is to serve only either a system of cold or a system of hot water; however, this volume should be at least 200 litres, ideally 250 litres if it is to serve both cold and hot supplies. The new cistern is installed as shown in figure 1.5 and as described in Chapter 06. In order to provide the best pressure possible at the outlet points, such as showers and taps, the cistern should be located as high as possible, which may require the construction of a supporting frame. If you make a stand it is essential that you use sufficiently strong timbers and brace it well to ensure it can take the weight of the cistern when full of water. The weight of water is quite substantial – 1 litre of water weighs 1 kg, so 250 litres weighs 250 kg (a quarter of a tonne)!

One point to note is that it is essential that no jointing pastes or compounds are used to make the connections to the cistern as these will have a detrimental effect on the plastic walls of the cistern, causing it to break down and reducing the expected lifespan of the new cistern. The connections to the cistern are made with what are referred to as tank connectors. These are simply passed through a hole made in the cistern, with a plastic washer included, and when the fitting nut is tightened it clamps tightly to the cistern wall.

Loft hatch too small for a replacement cistern?

Often the old cistern will have been installed during the construction of the building, when the roof was open. This may mean that the loft hatch is too small for a new cistern to pass through. This can pose a problem and in some cases will require the hatch to be made bigger. However, it is sometimes possible to buy a round cistern, the sides of which can be folded in, making it sufficiently small to pass through the opening to the loft. Alternatively, you can purchase two smaller cisterns and couple these together to provide an adequate volume, as shown below. Note that the overflow is in the same cistern as the float-operated valve, and the outlet is taken from the second cistern.

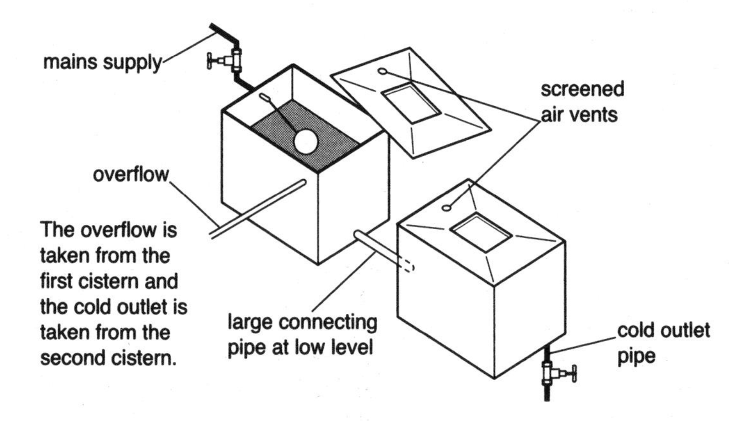

figure 8.4 coupling two cisterns together

Faulty immersion heaters

Should the heater element in an immersion heater break down and fail to operate it is a relatively simple process to replace it with a new one. The immersion heater consists of two parts:

- the heater element
- the thermostat.

If you plan to undertake any work on the heater, either to test the unit or to replace it, you must first isolate the electrical power supply to the heater by isolating the circuit and removing the fuse. Once you have confirmed that the power is dead you can:

1 Check the condition of the immersion heater by checking for a resistance in ohms using of a suitable multimetre between the line (phase) and neutral heater element terminals (see figure 2.7) The reading would be typically around 18 ohms for a good element, so a reading of 0 ohms would indicate a breakdown of the unit.
2 Check that the thermostat is operating properly. To do this, remove it from its pocket and connect a continuity tester to each side of the electrical connections. This is a function on a multimetre that bleeps when the two probes are put together. It should bleep, indicating a make or break connection when placed in very hot or very cold water. If the thermostat is faulty simply replace it with a new one.

Where the heater element is faulty a little more work is required, as follows:

1 Disconnect the wires from the terminals.
2 Close the water supply valve to the hot water cylinder. This is located on the pipe feeding the cold water to the cylinder (see figure 2.6). With the water supply isolated, open a hot tap fed from the cylinder and wait until the water stops flowing – this may take a minute or so. Now open the drain-off cock located at the base of the hot water storage cylinder to remove some of the water contained within; although no more water is serving the cylinder it is still a large container full of potentially very hot water. You need to drain sufficient water from the vessel to reduce the water level below that of the immersion heater. Where the immersion heater is located in the top dome of the cylinder, which is quite common, it is only necessary to drain off about 4½ litres of water (1 gallon); however, where it is located at some distance down the side of the cylinder it may be necessary to completely drain all of the water from it.

3 Once you have removed the water you can unwind the old immersion heater from its connection by turning the large nut anticlockwise. The spanner used for this is quite specialist and will need to be acquired from a plumber's merchant. Often the old immersion heater is held in quite solidly, in which case you will need to take a thin hacksaw blade and cut out the fibre washer which makes the seal between the immersion heater and the hot storage vessel. With this washer removed the nut will usually now unwind; if not, try turning clockwise to tighten it a little, thus breaking the seal. If it is still tight a little penetrating oil may be required to soak into the thread, or the heat from a blowlamp may provide sufficient expansion to effect removal.
4 With the old heater ELEMENT removed a replacement can be made, installing everything in reverse order and making sure you include a new fibre washer, smeared with a little jointing paste (see **appendix 2: glossary**).
5 With the new heater in position and tested for water soundness, the wires can be reconnected to the new thermostat.
6 Finally, the temperature on the thermostat needs to be adjusted to provide 60°C at the top of the cylinder.

Insulating to prevent freezing and frost damage

When water freezes it expands by 10 per cent. This expansion cannot be restrained and as a result it will cause the pipe or fitting to stretch, often to the point where it splits open. When the pipework splits open no water will come out as the ice will still be solid; it is only when the ice thaws that the problems start.

Should you need to repair pipework that has been subjected to frost damage it is worth bearing in mind that the whole section may have expanded prior to the split occurring. As a result, you may encounter difficulties in getting the new fittings to fit onto the pipe, due to its increased diameter, so you may need to cut out more pipework than anticipated. Fixing a burst pipe is like closing the stable door after the horse has bolted; the best thing is always to try and prevent the problem by insulating any pipework that could be exposed to damage in this way.

Insulating pipework is a relatively simple process and can generally be undertaken by anyone. You will find insulation materials for a whole range of situations at your local plumber's merchant. The obvious, yet foolish, thing to do would be to select the thin, relatively cheap material. The thinner insulation products are not necessarily designed for frost protection, although is could be said that they are better than nothing.

Insulation material serves several purposes:

- to provide thermal insulation against frost damage
- to prevent the loss of heat from hot water pipes
- to conserve fuel or to prevent heat loss from a domestic hot water pipe to a draw off point
- to cut down the transmission of noise to the adjoining structure, such as when installing pipework within internal timber stud walls.

Note: Should you have a major leak in your roof, causing all of the water to discharge through the ceiling, causing £1000's worth of damage, your insurance company may not pay out simply because your insulation material was insufficient, as laid down in the Water Regulations. The effectiveness of the insulation material is identified by the supplier. A suggested minimum wall thickness of at least 22 mm is advisable where flexible foam is used for internal applications, increasing to 27 mm for outside applications. Where loose fill materials are used this distance is further increased to provide a typical surrounding to the pipe of at least 100 mm.

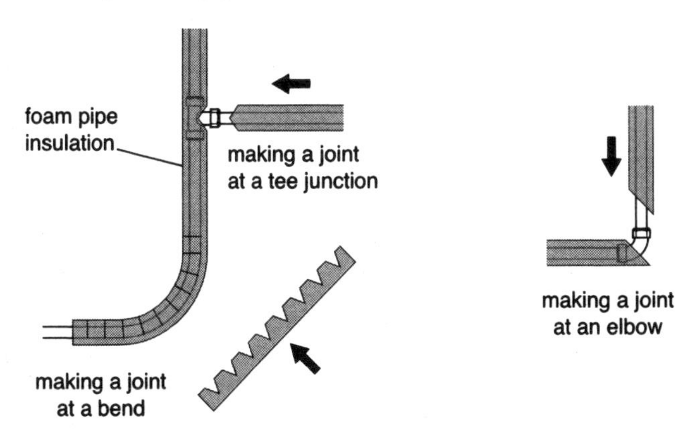

figure 8.5 insulating pipes with foam

If you look at figure 1.5 you will notice how all the pipework within a roof space has been insulated, including the vent pipe and overflow pipes. These will generally not have any water within them, but they may do if there is a fault, so these must also be fully insulated. In addition, the cold storage cistern itself must be insulated, apart from its base where it sits directly on the ceiling joists as hot air will rise from the building below. It is essential that all insulation material is securely fixed to prevent it opening and letting in cold air.

If you have another look at figure 1.5 you will notice that a bend is attached to the inside of the cistern overflow pipe connection. This bend turns down and should be dipping into the water. Its purpose is to prevent cold draughts from blowing up the overflow pipe and causing freezing conditions to occur inside the cistern. This has been a Water Regulations requirement now for over ten years but unfortunately the bend is all too often discarded by the plumber, or the water level is adjusted too low and below its inlet point, making this additional frost precaution void.

Installing new guttering and rainwater pipes

Old metal guttering systems are still obtainable if you need them. These are put together using nuts and bolts and a non-setting putty, such as 'plumber's mait'. Normal putty goes hard and therefore restricts the movement of the joints due to expansion and contraction caused by the heat of the sun, so should not be used. Plastic guttering is generally installed these days and is relatively easy to install.

The novice, when installing guttering, often makes two fundamental mistakes.

- First, they assume that the gutter requires a fall (slope towards the outlet) in order to remove the water from the channel. This is not so and in fact I often install guttering with barely any fall whatsoever; manufacturers' instructions suggest a fall of around 1 mm in every 600 mm. So, over the length of a building, say 9 m long, the total amount of fall would be 9000 mm ÷ 600 mm = 15 mm. This 15 mm drop will be hardly noticeable when viewed from ground level, so the guttering will look level and be more pleasing to the eye.

If you run a gutter with a greater fall than this the angle it is at will look strange. Also, the run off towards the lower end and outlet will be so great that water running along the channel will most probably discharge over the top of the gutter stop end.

- The second error is allowing insufficient room for expansion and contraction. Plastic expands quite extensively due to temperature change; for example, the 9 m gutter mentioned above, when subjected to a temperature change of say 35°C, which is quite probable when you compare winter temperatures with those of summer, would expand by 57 mm. This may not sound a lot but, if it is not allowed for, the guttering will buckle and a clip might break. Conversely, if the gutter contracts by this amount due to cooling it could cause a joint to be pulled apart. If you look carefully at a gutter fitting you will notice that the manufacturer imprints a line within the moulding to tell you where to finish the gutter end.

So, to install a new system of guttering:

1. Remove any old guttering materials, if applicable.
2. Review the condition of the timber facia board, i.e. the timber onto which the gutter is to be fixed. Now would be a good time to undertake any repairs and paint work to this.
3. Ascertain whether the facia board is level. A long spirit level should be acceptable for this. If you find that the facia is not level you will need to take particular care with the individual gutter fixings, using the level at each fixing clip to ensure you do not position them wrongly. If the facia is level you can simply fix one clip towards one end of the facia as high as you can.
4. Do the following calculation: total length of the gutter in mm ÷ 600 = total fall in mm.
5. Fix the next clip towards the other end of the facia at the calculated fall in height.
6. With these two clips fixed you can now tie a length of string between the two to act as a guide for fixing all the remaining clips and gutter fixings, such as connection joints and the outlet.
7. All fixings should be spaced at a maximum of 1 m apart (see figure 8.6). With the clips and fixings in position all it leaves is for the gutter to be snapped into place, ensuring the gutter has been cut to the correct length, where applicable, allowing for the expansion as identified above.

Although safe working practices are not within the remit of this book, I would like to mention a point of safety here. Remember that when working at heights you must take certain precautions, such as having well-supported ladders. It is ill advised to undertake this sort of work unless you have some understanding of the possible dangers and how to avoid them, in which case it may not be worth the risk of doing it yourself. So, if in doubt get a professional to do the work.

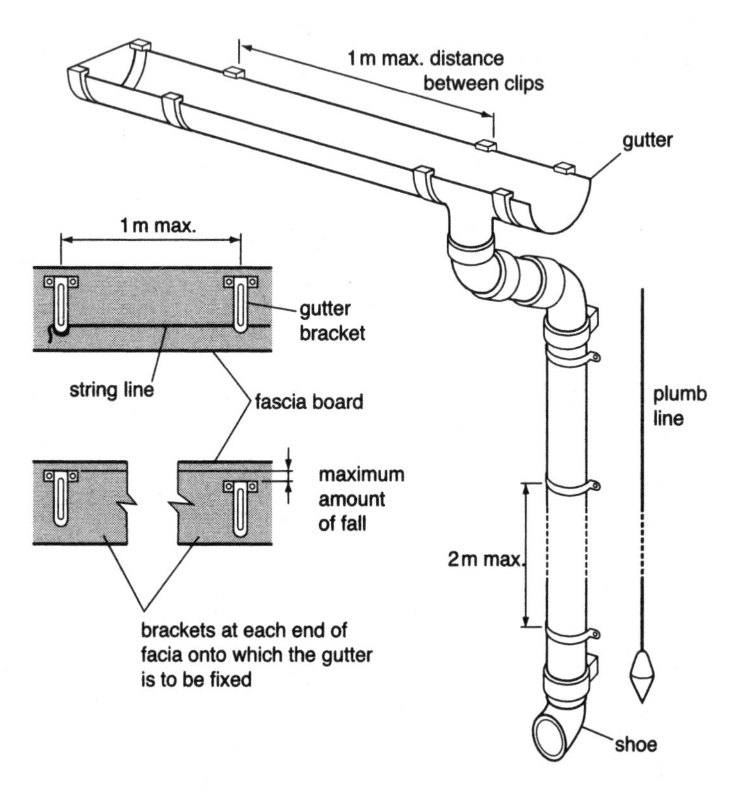

figure 8.6 installing gutters and rainwater pipes

With the gutter in place you can now run the rainwater pipe. This either terminates at its lowest end with a special outlet referred to as a rainwater shoe, which discharges into an open gully, or it is run into a drainage connection located at ground level. When installing this down pipe, as is sometimes called, it should be run vertically, again for reasons of appearance. This can be judged by placing the pipe clips at a maximum of 2 m apart in line with the brickwork or the corner of the building. Sometimes installers use a plumb line for this. This is basically a weight tied to a length of string and hung from the gutter outlet, to drop vertically downwards.

Remember when fixing the clips, and in particular the coupling or joining sockets, to make sure an allowance has been made for the expansion of the pipe to avoid it buckling due to temperature changes.

When assembling the gutter and rainwater pipe, no additional jointing mediums are required. The guttering is fitted with a rubber seal incorporated into the gutter fitting. The rainwater pipes require nothing as the joints just slot together with the pipe end dropping downwards into the socket of the fitting below.

Installing a new WC suite

The installation guidance given here will assume the worst and deal with the larger job of converting, say, a high- or low-level suite to a combination suite (a suite where the cistern sits directly on the pan and does not have a flush pipe). Obviously changing a like-for-like installation is a much simpler task but will follow similar guidelines.

Removal of the old suite:

1 Turn off the water supply.
2 With the supply isolated, give the cistern a final flush to remove most of its water. Then remove it from the wall, tipping any remaining water into the pan.
3 Next, remove the toilet pan. If it is no longer wanted, you do not need to take any particular care of the sanitary ware you are removing (the pan or the cistern), so if you need to put the hammer through the foot of the pan because it is cemented to the floor or to the drainage pipe then do so. Don't forget, however, that there is a little water contained within the trap that needs to be disposed of first.
4 With the old suite removed you now need to cut back the water supply pipe to a position from which it will be suitable to run to the new WC cistern.
5 I would now suggest fitting a temporary cap or blanking fitting to the water supply pipe, allowing you to turn the water back on until the work is near to completion, as it may be off for some time.

All going well, the work you have done so far should have taken no more than an hour. You now need to install the components of the new flushing cistern into their respective positions within the cistern. This is a relatively simple operation and the manufacturer will have supplied some installation instructions.

6 Note that you will need to change the size of the inlet seating of the float-operated valve or insert a special restrictor, depending upon whether the water supply is on high or low pressure. Again, the installation instructions will give advice on this.
7 With the components assembled you can now carefully put the WC pan up to the position where it is to be located and hold the cistern in position to check that everything will fit. Unfortunately there are sometimes problems with this.

- First, the pan outlet is sometimes either slightly too high or too low and does not align correctly with the soil pipe branch connection.
  - Where it is above the soil pipe connection this is not generally a big problem, as you just need to purchase a 'Multiquick' offset pan connector, as shown in figure 8.7, to allow for the drop.
  - However, if you pan outlet is lower than the soil pipe connection you do have a problem as simply using an offset pan connector will create an additional uphill obstruction that the flush will have to overcome and that may become an area prone to blockage. So you need somehow to alter the soil pipe branch connection to lower it (this is a major job and involves dismantling the drainage stack), or you could put the pan on a hardwood plinth (clearly not an ideal solution). Fortunately this is not a common problem and it is likely that the old pan was located on a similar mounting in any case. The outlet height for all British designed WC pans is identical.
- A second problem sometimes encountered is that when you put the pan and cistern up to the wall the outlet pipe is too short and the pan outlet does not reach the soil pipe. This problem is easily overcome by using a WC pan extension or a 'Multiquick' extension piece. This can be inserted into the soil pipe and adjusted or cut as appropriate.
- Conversely, the soil pipe branch may be too far forward from the wall. This prevents the cistern from touching the wall. Where the cistern is held off the wall you will need to cut the

soil pipe back as necessary. This is not a problem with plastic soil pipes but where an old cast iron stack has been installed this may prove difficult without specialist equipment. In this case you may have to attach battens onto the wall for the cistern to sit against, again not the ideal solution.

Assuming that everything fits correctly:

8. Insert a plastic push-fit pan connector, such as the 'Multiquick', into to soil pipe and push the pan into this.
9. Now place a spirit level on the pan to check that it is level, packing up the sides if necessary, and screw the pan to the floor using 65 mm long brass screws. Sometimes pans are secured to concrete floors with sand and cement but this is not ideal as it prevents the pan from being removed in the future.
10. With the pan firmly secured in position, put the cistern up to the pan and mark out the location of the wall fixings.
11. Drill the holes as necessary and insert the wall fixings.
12. The cistern can now be securely bolted onto the pan, incorporating the foam donut washer supplied to form a seal between the cistern and the pan. The cistern can be secured to the wall using brass or stainless steel screws to prevent them from rusting.
13. The old cistern probably had an overflow connection run to the outside, but these days the overflow is generally incorporated within the cistern flushing mechanism.
14. The cold supply pipe can now be run to the cistern from the point where the supply was capped off. Within this pipeline you must incorporate an inline isolation valve just before the connection to the cistern, thereby complying with the Water Supply Regulations. This valve is generally of the small screwdriver-operated quarter-turn type. Take particular care when making the connection to the plastic thread of the float-operated valve as it is very easy to cross-thread with the brass tap connector used against this soft plastic material.
15. Now turn off the water supply, remove the temporary cap and make the final water connection.
16. Turn on the water supply, adjust the water level as necessary (as indicated by a line inside the cistern) by setting the position at which the float shuts off the water supply and test out the installation.
17. Finally, if you have not already done so, fit the toilet seat to complete the task.

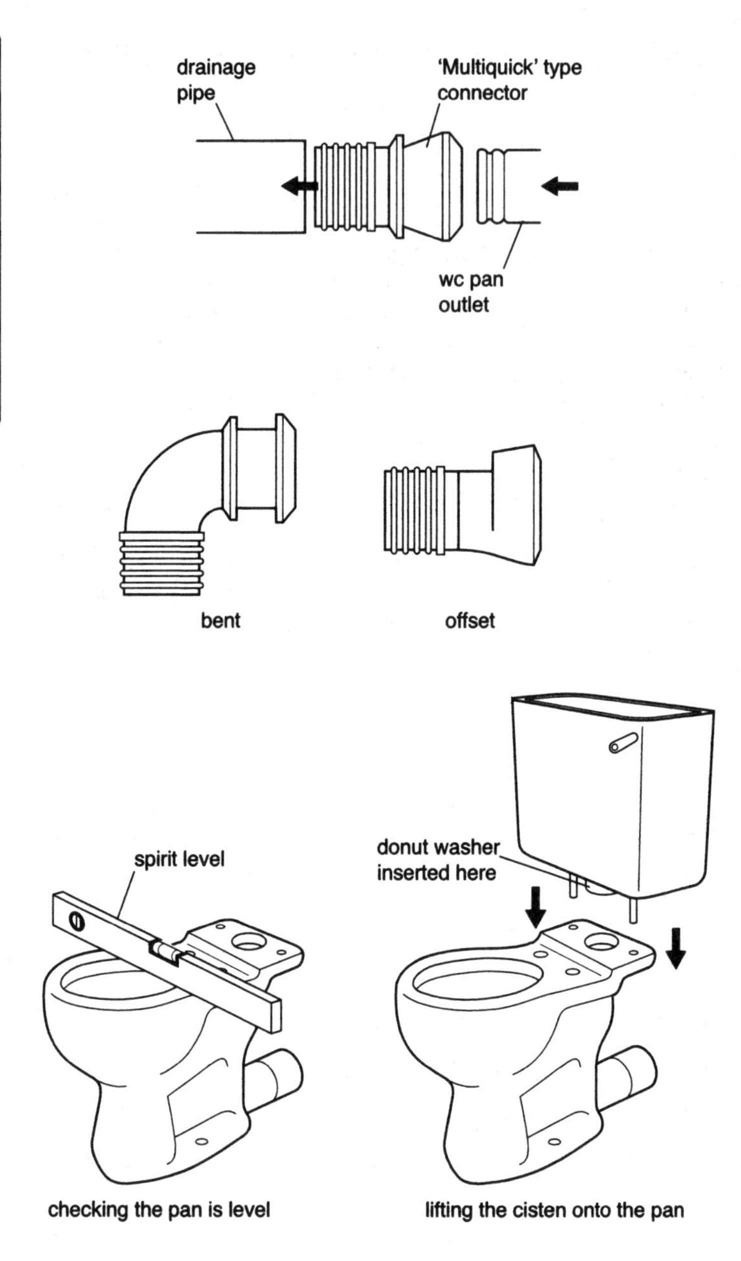

figure 8.7 installing a close-coupled WC suite

Installing new sink, basin or bath taps

If you plan to change your appliance taps, possibly with something more modern, the first thing you need to do before purchasing your new taps is look at your existing hot and cold water supplies. Where you plan to install a mixer tap you must determine whether or not they are being supplied by pressures that are equal. If you refer to figure 8.8 you will notice that one mixer tap is designed to mix the water within the body of the tap and this should only be used where the pressure to the hot and cold supply to the tap is the same. The other tap design does not allow the hot and cold water to mix together until it leaves the spout, so it is okay if they are of different pressures.

It is essential to recognize that where the water pressures are different (such as a high-pressure cold mains supply and a low-pressure, cylinder-fed hot water supply), if you use the wrong tap then when both are opened at the same time it will not be possible to get water from the low-pressure pipework. The high-pressure water will take precedence and could back up via the opened tap to pass into the other lower-pressure pipework. *So your need to choose the right design of mixer tap!*

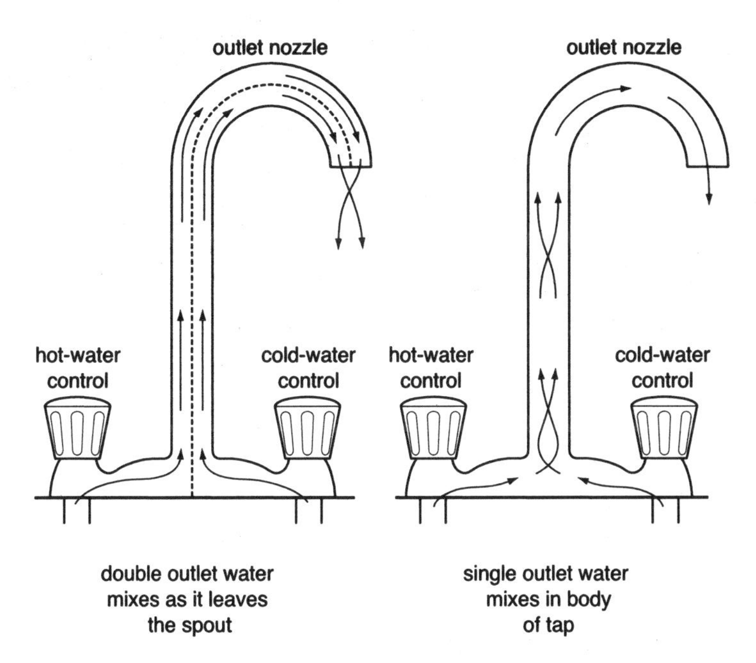

figure 8.8 single and double outlet mixer taps

Some of the fancy modern taps are designed to give a frothy or pulsating discharge from the spout. You may need to consider the flow rate and pressure required in order to get this effect and whether your existing supply can meet these requirements. If not, the taps will not perform properly. However, for standard taps this should not be a problem.

Once you have purchased your new tap/s, before beginning the work you may need to alter the pipework a little as the length of the thread on your new taps may differ from the existing fitment. So, don't start the work when the shops are closed as you may need to buy a few pipe fittings! In order to complete this task you will also need to purchase a basin spanner, a specialist plumbing tool available at most plumber's merchants (see figure 6.9).

When you are ready to start, the taps are changed as follows:

1 Turn off the water supply to the tap/s and confirm that the water has stopped flowing. Leave the taps open and then, if possible, open another tap or flush the toilet on the same section of pipework that has been turned off. If you watch the lower of the two open valve outlets, assumingly your tap is the lower of the two, you will see that it starts to flow again, possibly discharging a significant amount of water. This is because as the other appliance opens, it lets air into the pipework to allow the water to escape. If you do not open this second appliance you will have a continued slow discharge as the air slowly enters the pipe. This generally runs down your arm as you are lying on your back reaching up to the tap. Or, failing that, when you are working on the job someone in the building might open another tap or flush the toilet and you will get completely soaked.
2 Once you have gained access to the underside of the tap connection you can use a basin spanner, turning it anticlockwise, to undo the tap connector which joins the tap onto the pipework.
3 You now repeat the process and undo the back nut, which clamps the tap to the appliance. You will need to get someone to hold the tap still to prevent it from turning while undoing this nut.
4 You should now be able to remove the old tap, unless it is a mixer tap, in which case you repeat the process for the other water supply.

5 Now that you have removed the old tap you will see if the pipework needs to be altered to allow the new tap to be installed. It is hoped not, as invariably there is sufficient movement within the pipework to accommodate a slight difference in the length of the thread. Sometimes, where a new tap is a bit short, it is possible to include a shank extender/adapter to provide the additional length required. Alternatively, you may need to alter the pipework, in which case a bendable pipe tap connector may prove useful for those not too proficient at bending pipes. If you need to alter the pipe work refer to the basic plumbing processes in Chapter 06.
6 The new tap can now be inserted into the appliance with the supplied foam or rubber washer, or with a ring of plumber's mait around the base of the tap as it sits into the appliance. This forms a seal to prevent any water that has splashed onto the appliance passing through the gap and dripping onto the floor.
7 Firmly secure the tap into the appliance with its back nut.
8 In the case of stainless steel sinks where the appliance is made of a very thin material, a spacer washer is required, sometimes referred to as a top hat. This is positioned between the appliance and the back nut, making up the gap where there is no thread on the top part of the tap shank.
9 Now wind the tap connector onto the threaded shank of the tap and include a new fibre washer to form a good seal.
10 Finally, turn on the water to test out the installation. If there is a slight drip from the tap connector you could try tightening this fixing a little, but often after a few moments it seals itself as the fibre washer becomes wet and expands.

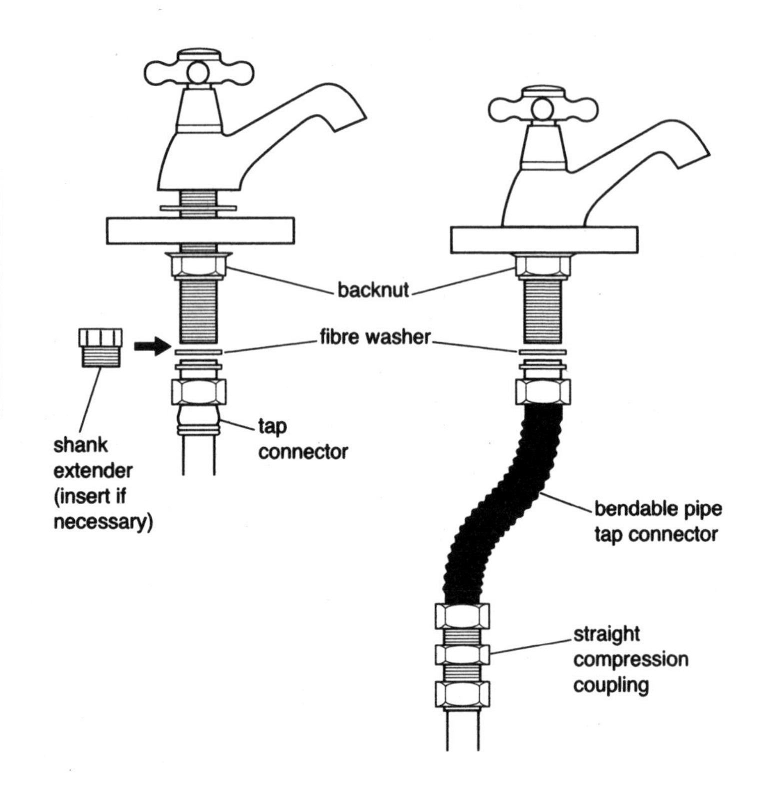

figure 8.9 tap connections

Installing a new sink, basin or bath

It is likely that if you are installing a new appliance it will include a new set of taps, in which case you can refer to the notes above to ensure that these are suitable for the installation in terms of pressure, flow and mixing.

The new installation may be as the replacement for an existing appliance or it could be a totally new installation. If it is as a replacement it is possible that the existing services could be used without alteration, including the water supplies and drainage connection. These notes assume that this pipework has to be installed, but if you just need to make an alteration, ensure that you do not cut back the existing pipework more than is necessary.

For a completely new installation the first thing to do is to check that there is a clear route to run the waste pipe to a drain. It is generally possible to run the water supplies anywhere, but the waste pipe requires a gradual fall in the direction of the drain, where the connection can be made.

Look outside the building for a gully or hopper head (see **appendix 2: glossary**) into which you could discharge your new waste pipe. Alternatively, you may need to make a boss connection into the soil stack as it passes down to ground level. This is a much bigger job and for this you will need to see the section above on making a branch connection to the existing vertical soil pipe.

Where no drain connection is viable it may be possible to include a macerator pump, in which case you should refer to the notes in Chapter 01. Ideally, the pipework should to be installed following the guidelines given in Chapter 01, with a minimum waste pipe size of 40 mm (1½") for baths and sinks and 32 mm (1½") for basins. However, if your pipe length exceeds these requirements it is possible to install a slightly larger pipe and reduce it in size as necessary as it gets nearer the appliance. Alternatively, a resealing trap could be considered (see figure 1.16).

1 So, the first job is to run your waste pipework and terminate this at a suitable location beneath the proposed location for the new appliance. This needs to be at a position no higher than the final location of the appliance trap outlet (see below). Also, where a pedestal basin is to be installed it is best to keep the pipes behind the pedestal for reasons of neatness, and this should also be considered when deciding where to terminate the waste pipe.
2 You can now run the hot and cold water pipes from where they are to be connected to the system to a location beneath the new appliance in a similar location to the waste pipe. See Chapter 06 for advice on running pipework and making joints. No connection to the water supply should be made at this time.
3 Running the waste pipe and water supplies in this way is referred to in the trade as the 'first fix' pipework, i.e. the pipework has been run to an installation, but the appliances have yet to be fixed.
4 The appliance is now made ready, which means that the waste outlet connection is secured to the appliance and the taps secured as necessary, doing up the back nuts clamping them to the appliance. Making the waste fitting into a basin is different from that of a sink or bath, as explained below.

Bath and sink waste fittings

These use a thick rubber washer beneath the appliance outlet, and possibly a second thinner washer above. These are clamped together with the appliance in between, with the aid of a large back nut, although some designs use a long stainless steel screw (as shown in figure 8.10), clamping the waste tightly to the appliance. The overflow is connected to this waste fitting with a flexible pipe which is likewise securely clamped to the appliance overflow hole.

Basin waste fitting

For basins there is no need to make an overflow connection as this forms an integral part of the appliance. The waste fitting is ideally made into the basin with suitable rubber washers but where these are not available the joint can be made as follows:

1. Apply a ring of plumber's mait or silicone rubber to the underside of the section of the waste fitting that sits in the basin outlet. For this joint to be successful the appliance must be absolutely dry, otherwise the jointing mediums used will fail to stick to the porcelain.
2. Place the waste fitting in position and apply a second ring of plumber's mait, silicone or a large rubber washer to the area around where the thread pokes through the waste hole of the appliance.
3. Then put on a 32 mm (1½") polythene washer.
4. Finally, wind a large back nut onto the waste fitting, clamping the whole lot together to form a seal.
5. To prevent the waste fitting turning in the basin while doing up the back nut, poke two screwdrivers through the slots or grates in the waste fitting and hold it secure.

Once the appliance is made up you can begin what is referred to in the trade as the 'second fix'. First, secure the appliance into its location, ensuring it is adequately supported and level. The top of the appliance is plumbed in level as the gradient towards the waste is built into the design of the appliance.

For plastic bath installations there are additional wall fixing clips, identified on the instructions provided with the appliance, designed to prevent the bath sagging due to the weight of the water as it fills. Also, if a bathtub is to be located along a wall that is going to be tiled, and the edge of the bathtub has a widely curved edge for which the depth of the tiles does not provide suitable coverage across the top, you may need to cut a chase

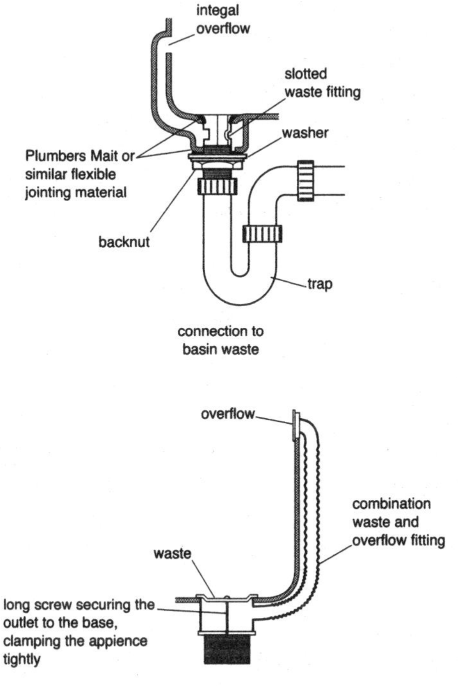

figure 8.10 connections to a basin waste fitting

into the wall to allow for this. Failure to do this means water will accumulate at this ledge and cause staining, and may lead to a leak past this point onto the floor.

Making the waste connection

The final connection to the waste pipe can now be made by installing a suitable trap onto the waste fitting of the appliance. With the waste connection finished you should get a bucket of water to give it an initial test because if you do need to remove the appliance again, for any reason, you don't want to undo more than is necessary. If a basin waste fitting is leaking where plumber's mait or silicone rubber have been used then unfortunately you will need to take it out, completely dry off the porcelain and remake the joint.

Making the water connection

Make the hot and cold water connections to the taps from the pipe previously located beneath the appliance. This can be a tricky process without the use of a bending machine, but special flexible connectors are now manufactured that help to make this joint and I would always recommend that you pay that little extra to make this final fiddly section of pipework. An example of one of these is shown in figure 8.9.

Once you have made these connections all that is left is to do is to make the final connections to the water supplies – remember that these have not yet been made. For this:

1 First you would need to turn off the water supply.
2 Confirm that the water has been shut off before cutting into the pipe to make your connection.
3 Finally, turn the water back on and test out your installation.

Your plumbing works are now finished, but there is one final requirement in relation to any metal components contained in a bathroom and this is to make sure that all the metal work is suitably bonded together, thereby ensuring it is all at the same electrical potential. This means that an earth wire needs to be joined to all the exposed metal parts, in effect linking them together. See the notes on equipotential earth bonding in Chapter 01.

The finishes can be now made good around the appliance, such as tiling and applying a fillet of silicone rubber along the top edge of the appliance to prevent water splashes dripping down behind it onto the floor below.

Installing a shower cubical

When you are considering installing a shower cubical, one of the first things to decide is how to gain access to the waste pipework at a later date should you need to, for example, to unblock the pipe. You can purchase traps that give access via a pocket above the appliance that lifts out from the waste fitting itself, but you can never guarantee that this will be sufficient. Once the shower tray is installed it will not come out again without causing a lot of damage, therefore trap access is essential. Sometimes it is possible to access a trap fitted to the appliance from a floorboard access door next to the shower tray; if not you will need to consider:

- an access panel located in the ceiling of the room below the shower cubical
- raising the shower cubical onto a plinth, in which an access door has been fitted

- installing a running trap, such as that shown in figure 1.12 underneath the floorboards within an area just outside the appliance area.

With this decision made the waste pipe can be run in the same way as for a new sink, basin or bath, as outlined above. The size of a shower waste pipe should be no less than 40 mm, i.e. the same as a bath waste pipe. The shower tray now needs to be secured in position, again following the guidance for fitting baths if there will be tiling along the edge of the shower tray, it might need the tray to be recessed into the wall. You must follow the manufacturer's installation instructions for securing the shower tray in position to avoid any movement and ensuring it is level along the top edge in all directions. The fall to the waste outlet is built into the appliance.

With the waste connection completed and the tray adequately secured you can now consider the water connections. For this you may choose to use:

- a cold water, mains-fed shower heater
- a combination boiler or multipoint water heater
- a storage-fed hot water supply, with or without a booster pump.

There are other options, but the above cover most scenarios.

The cold water, mains-fed shower heater units only require a cold water supply, usually supplied under mains pressure and with a minimal flow rate, therefore you will need to check the requirements for installation with the supplier of the heater. These units do require an electrical supply to be installed and this must be on its own circuit from the electrical consumer unit or fuse box. This electrical supply will also need to be certificated and the work is notifiable to the local Building Control department. Basically, the water supply is run to the small heater installed within the shower cubical and the shower hose is simply connected to the heater. The electrical power is initiated by the operation of a pull cord adjacent to the shower itself.

If the second option is chosen and the water is taken from an instantaneous system such as a combination boiler or multipoint, you must remember that the water flow is invariably restricted as it flows through the heater and therefore water volumes may be restricted, adversely affecting the performance of the shower.

With both of these options, where the cold water to a shower is taken directly from the cold water supply mains you must ensure that no backflow of water can occur, i.e. no water can be sucked back into the water company's mains supply. This is usually achieved by ensuring that the shower head cannot be laid down at a point below where it could be submerged in water.

If you choose a storage-fed hot water supply option for your shower, note that in order to ensure that the cold water is never starved from the shower mixing control valve it is taken independently from the cold water storage cistern as its own supply. The cold supply is also taken at a location within the storage cistern below that of the cold feed to the hot water cylinder. In both cases, this is to ensure that you can never stand in a flowing shower that goes very hot because the cold water has ceased to flow, i.e. the cold water will always be the last supply to stop running and it therefore prevents scalding.

Ideally the shower should also be supplied with the hot water taken directly from the hot water storage cylinder, independently from the other hot water draw-off points. To ensure that the shower works effectively, the water pressures from both the hot and cold water need to be the same (as identified above in relation to mixer taps). When connecting to a storage cistern you must ensure that you have at least 1 m head above the shower rose to the underside of the water cistern in the roof space, otherwise the water flowing from it will be very poor.

An alternative is to incorporate a shower booster pump. These are very effective but you must follow the manufacturer's guidance and maintain a good flow rate of water to the booster. This usually requires a minimum of 28 mm diameter pipework, otherwise its lifespan will be reduced as it struggles to cope with the limited water supply. Domestic shower boosters are very simple units to install, usually coming with push-fit flexible water connections and a pre-wired 13 amp plug that is simply plugged into a nearby socket outlet. They automatically operate as the water flows through the booster, initiated by an internal flow switch, hence the need for an adequate supply.

In all of the cases above, the pipework to the shower control should be run following the earlier guidance notes on running the hot and cold water pipework to various appliances. The actual shower mixer could be installed onto the finished tiled surface or built into the wall, in which case all the pipework will also need to be located within the wall and the whole system tested prior to tiling.

Finally, where metal pipework or components have been installed these need to be suitably bonded (see Chapter 01).

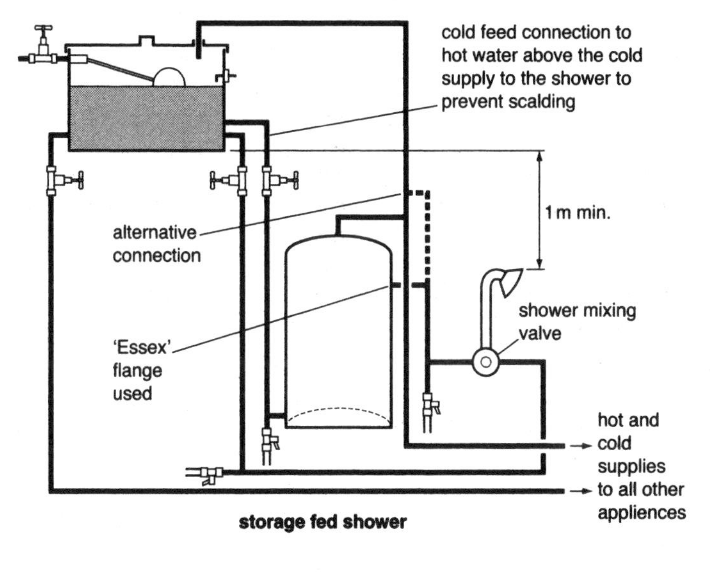

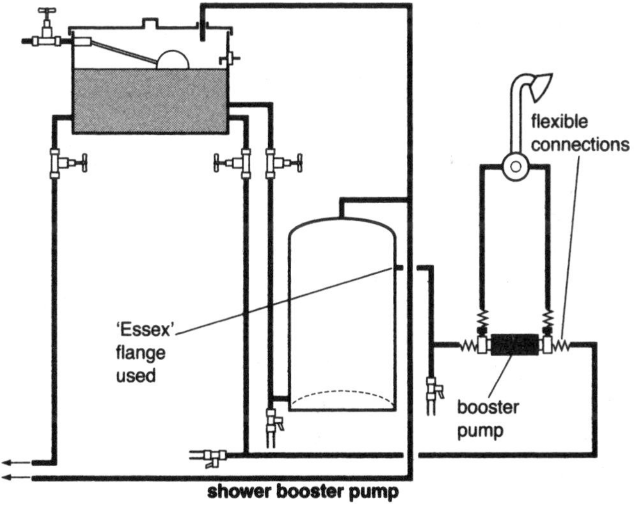

figure 8.11 installation of a shower

Replacing a shower booster pump

This is a very simple operation. You will need to purchase a replacement pump with identical or similar qualities. The electrical connection is generally a pre-wired 13 amp three-pin plug that has simply been plugged into a convenient socket outlet, therefore there should be no problems with disconnection and reconnection. Once you have everything at hand the task is completed as follows:

1 First, isolate the hot and cold water to the unit and confirm that it is off.
2 If available, open the drain-off cocks to release any water from the pipework feeding the pump. Alternatively, be prepared for a little water to discharge as you disconnect the pipework. Fortunately these connections are usually made via a flexible connection with push-fit or compression joints, your new replacement booster will almost certainly have the same sort of connections. So it is often simply a case of removing the old connections and fixing on the new ones, as the flexible connections will allow for the necessary free movement to facilitate the replacement.
3 Turn on the water supplies and check for any leaks.
4 Plug the new pump into the socket outlet provided for the existing pump.

Where no flexible connections have been provided you will have to be prepared to alter the pipework as necessary, following the notes in Chapter 06, taking care not to damage the plastic water connections of the new pump with heat from any soldering processes or cross-threading the connection.

I could give general instructions for installing endless appliances but, as mentioned at the beginning of this chapter, all jobs follow the same basic principles. You just need to transfer the skills outlined here. So, in conclusion:

1 Turn off the water supply and hot water heaters where applicable.
2 Confirm that the water has shut off correctly.
3 Remove items that are no longer wanted.
4 Temporarily cap off the pipework at a point where the new connection is to be made, thus allowing you to turn back on the supplies until the new installation is ready.

5 Prepare the work area for the new installation, running all new pipework and installing the appliances.
6 Turn off the water supply and make the new connection as necessary.
7 Turn on the water and test out the installation as appropriate.

appendix 1: legislation

Gone are the days when everyone could do what they liked. Today there is a whole range of legislation effecting what we can and cannot do in our homes. There is no restriction on what you can do yourself, but you would still need to ensure that your work is in compliance with the law.

Much of the work completed these days requires the issue of a completion certificate, which is something, incidentally, which you must insist upon when employing someone else to do the work for you.

- Do not assume they are registered with a specific body (see below).
- Do not be fobbed off with 'its not applicable to what they are doing'.

Most activities completed these days require some form of certification. The following gives a guide to situations requiring certification by the local building or water authority. When you have the work done you may not care too much whether or not a certificate is issued, but:

- when you come to sell your home it may be picked up by the surveyor and prove costly to certify this work at a later date
- you may not be covered should you wish to make an insurance claim.

Work requiring notification under local Building Control

In *all* of the following situations you will need to notify the Local Building Control Officer of the local authority of work carried out. This is in order to comply with the requirements of the Building Regulations.

Where you use a contractor to do the work they may be registered with a validating body such as CORGI[1], NAPIT[2], APHC[3] or OFTEC[4], which allows them to self-certify the work (note that there are other bodies). However, you need to check that they are registered with a certificating organization or you may not receive the certificate for the work completed, as required by law.

Drainage alterations

Certification is required for all new additions to your drainage system, such as an additional toilet, sink, bathroom or pumped macerator unit. It is also required in all instances where you wish to alter your existing waste pipework, for example if you want to move your bath from one corner of the room to the other. The only time when notification is not required is where you do not alter the waste pipework at all and use the existing connections for a straightforward replacement.

Heating and hot water requirements

Where a new boiler or hot water cylinder is to be installed, notification and certification will be required. When just a new cylinder is needed it is sufficient to undertake the appropriate replacement and ensure boiler interlock is provided for (see Chapter 03), but where the boiler is to be replaced it must be upgraded, in most cases to a high-efficiency type, often referred to as a condensing boiler. In addition, the heating system will need to be upgraded to include all of the items listed under the heading 'Heating controls' in Chapter 03.

The installation of unvented domestic hot water systems also requires notification.

[1] Council of Registered Gas Installers (CORGI: 0870 401 2300)

[2] National Association for Professional Inspectors and Testers (NAPIT: 0870 4441392)

[3] Association of Plumbing & Heating Contractors (APHC: 024 7647 0626)

[4] Oil Firing Technical Association (OFTEC: 0845 65 85 080)

Electrical systems

Alterations to the electrical installation are subject to certification. Additional alterations in rooms such as general living areas and bedrooms do not require notification to Building Control, but they will still require an electrical minor works certificate to be issued. Areas that must be reported to the local authority include new circuits or when work is completed in wet areas such as bathrooms/kitchens and where a central heating control system has been installed.

Gas systems

Where a new gas appliance is installed, including cookers, gas fires and boilers, the work must be certified. The only body currently registered to self-certificate the installation of a domestic gas appliance is CORGI.

Oil installations

Oil installations and replacement of oil boilers and storage tanks also require certification. Operatives able to self-certificate will be registered with OFTEC.

Ventilation

Where an extractor system is included, such as in a bathroom, notification is required.

The requirements for Building Control notification do not stop at the above areas of work. Where you alter a building, certification may be required, including changing your windows. You need to keep hold of these certificates and know where they are should you wish to move home.

A couple of points to note:

- Advice relating to any of the above situations can be sought simply and quickly by phoning your local council offices and asking to speak to the Building Control Officer. The plumber you engage should be fully conversant with these rules, but I suggest you do not bank on it!
- Without certification you may have to remove what you install if it is discovered by the local authority to have been carried out without approval.

Work to which Gas Regulations apply

You may undertake gas work as DIY in your own home but, if you do, it is essential for your own safety that you fully understand what you are doing, and your work must be in accordance with the gas safety regulations. If you decide to employ a professional to do the work, in addition to the certification discussed above, the installation of a new gas appliance and all gas work undertaken within your property must be undertaken by a CORGI registered gas fitting operative. It is very important that you ask to see proof of the operative's identity in relation to the work that they wish to undertake, as failure to do so might put your life as risk. All operatives have been issued with a CORGI yearly credit-card sized proof of identification. This card has their photograph clearly displayed on the front, which you must check. Their identity can be checked by phoning CORGI on 0870 401 2300 or by logging on to www.trustcorgi.com.

Do not just look at the front of the card. Listed on the back are the specific areas of domestic gas expertise in which the operative is authorized to undertake work; these include:

- pipework
- cookers
- gas fires
- water heaters
- central heating
- warm air heating
- tumble dryers.

So, for example, if you want work done on your boiler, make sure central heating is listed on the back of the card. If you find that an operative is operating without registration you should report them to CORGI as they may be endangering the lives of others. If they do not produce the card then do not let them do the work.

Work to which Water Regulations apply

Under the Water Regulations you have a duty in law to prevent the following situations occurring within your premises at all times:

- wasting water, e.g. not repairing leaking joints or dripping taps
- misusing water, e.g. filling a swimming pool in excess of 10,000 litres without giving notice to the supplier
- undue consumption of water, e.g. installing a toilet cistern of a larger capacity than that permitted
- water contamination, e.g. using lead soldered fittings instead of lead free fittings
- erroneous measurements of water used, e.g. bypassing a water meter.

As with all the other regulations, you can undertake any new work in your home yourself but you must complete the work in accordance with the regulations listed above. In addition to this, some specific new works or additions to your plumbing system will require notification to the local water authority; these include.

- erecting a new building – permission is required both for internal pipework and for the use of a temporary water supply for building purposes
- installing a bath with a capacity greater than 230 litres
- installing a bidet with an ascending spray or hose
- installing a booster pump using more than 12 litres per minute
- installing a water treatment unit such as a water softener
- installing a garden watering system, unless it is a handheld hosepipe
- constructing a pond or swimming pool greater than 10,000 litres filled by automatic means.

If you wish to undertake any of these activities as part of your new works you must apply in writing to the local water authority at least ten days before starting the work. They will either:

- give consent
- refuse consent, giving their reason
- give consent subject to certain conditions.

If you do not receive a reply after ten days, consent may be deemed to have been given.

When employing a plumbing contractor to do the work for you, you should check to see that they are approved and registered with a water authority. This will ensure that the work is in compliance with the Water Regulations. Also make sure they issue you with a certificate once the work is completed and certainly do not settle your account until you have this. As with Building Control certificates, these need to be kept and produced if called upon to show you have had approval.

Where you employ a plumber who is not registered, you may have difficulty proving compliance with the law and any certificate offered by them will be worthless. Also, if they are not approved you cannot begin the works listed above until consent has been given.

It must be understood that *you*, the householder, and therefore the user of the supply (not the person who actually did the work) will be held liable for contraventions to the water regulations and therefore you will be subject to any fines imposed due to contravention.

Conclusion

To recap:

- For all work relating to gas installations unless DIY, operatives *must* be registered with CORGI.
- For all work relating to oil,* operatives should be registered with OFTEC.
- For all work relating to drainage,* operatives should be registered with an approved body, such as those listed above or another such organization.
- For all work relating to the hot or cold water pipework,* operatives should be registered with a water authority.
- For all work relating to the electrical installation,* operatives should be registered with an approved body, such those listed above or another such organization.
- For all work relating to the ventilation,* operatives should be registered with an approved body, such those listed above or another such organization.

* Where the operative is not registered with a specific body the work can still be completed, but you may need to seek approval *in writing* first from the local building or water authority, as necessary.

The professional plumber

Finding the right plumber may prove difficult. You will be lucky to find someone with everything I have suggested a plumber should have. I have a lot of experience in training and meeting with plumbers, heating engineers, gas fitters and electricians. These operatives need to be registered to work on:

- gas installations and tested on each type of appliance they wish to work on
- hot and cold water systems
- drainage systems
- electricity supply systems
- heating systems.

Each of these disciplines requires:

- upgrade training
- passing an assessment, often on a five-yearly re-assessment basis
- paying the annual fees to belong to a professional body, of which they may need to join several.

All of this is very expensive!

In addition they have to:

- take time off from work to attend courses and therefore incur a loss of earnings
- comply with an extensive additional safety legalisation, which has additional cost implications.

All this has resulted in operatives being faced with the decision to either limit the work that they can do legally, or offset the cost of the work through the fees they charge for their work. The amount they quote for a job is therefore often above what the client had envisaged paying. It will almost certainly be higher than a quote from someone who is not appropriately registered to carry out the work.

Check why someone is asking a specific price for a job. They might be the most expensive because they will give you everything you need. You need to make sure they are the dearest for a reason and are not just fleecing you. Conversely, will the cheapest quote give you the standard of service you expect and be certified if applicable?

Good luck in your search. Find a good plumber and keep them.

appendix 2: glossary

access cover A point where access can be gained for internal inspection of a drain.

air gap A distance maintained above the top of an appliance, such a sink or basin, at which the water would spill over onto the floor to the underside of the tap outlet. An air gap is maintained to prevent back siphonage of water from the appliance into the supply pipe.

air separator Also referred to as a de-aerator, this device is sometimes found in a fully pumped central heating system to maintain the neutral point, where the cold feed and vent pipes join the system.

alloy A material made up from two or more metals, for example brass or solder.

automatic air-release valve A special valve that allows air to escape from any high points within a low-pressure system, where air would accumulate and cause a blockage.

back boiler The term generally given to a boiler that has been installed within the fire opening in a living room or lounge.

backflow Water that flows in the opposite direction to that intended, possibly causing water contamination.

back siphonage Water that is sucked back into the pipework, causing water contamination.

balanced flue See *room-sealed appliance.*

balancing The term used to indicated the throttling down of the flow of water to certain radiators to force the water to flow to those further away from the pump, thereby allowing them to receive some of the hot water.

ballvalve See *float-operated valve.*

basin spanner A special spanner designed to reach the nuts located up behind baths and sinks where there is restricted room to turn a normal spanner.

bib tap A design of tap that screws horizontally into the pipe fitting, such as used for an outside tap. See also *pillar tap*.

boiler The appliance that is used to warm up the water for washing and central heating purposes.

bonding A system where specific metal pipework is connected either together or to the main earth terminal in order to combat the potential of an electric shock arising from a faulty electrical installation.

boss A special connection branch where smaller diameter pipes are connected into a larger diameter waste pipe.

branch A term used to identify a tee joint.

brass An alloy of copper and zinc.

Building Regulations The laws applicable to building works. They are administered by the local authority.

capillary joint A soldered joint used to join two pieces of copper tube together.

carbon monoxide A poisonous gas produced as the result of incomplete combustion of fuel.

cesspool Also called a cesspit, this is a sewage collection chamber where foul drainage water is collected until it can be removed for proper disposal.

checkvalve A non-return valve fitted within a pipeline, designed to prevent backflow.

circulating pump A device that is installed in a central heating system to circulate the hot water to all the radiators.

cistern An open-topped vessel used to store water, such as a water storage vessel or a toilet cistern.

close-coupled suite A toilet suite that does not include a flushpipe. The cistern is bolted directly onto the WC pan.

cock A type of valve, such as a stopcock.

cold feed The pipe that serves a specific system, such as the cold feed to a hot water system or central heating system.

combination boiler A design of boiler that is used for the central heating and as a unit to instantaneously heat up water going to hot water outlets. It does away with the need to have a hot water storage cylinder.

combined system of drainage A system of drainage that takes both rainwater and the foul water from the house sanitary pipework.

compression joint A fitting used to join two pipes together. It incorporates a compression ring that is clamped onto the pipe and forced up hard against the body of the fitting.

condense pipe A pipe that is used to remove to a drain the water collected within a condensing boiler.

condensing boiler A design of boiler that operates to a very high level of efficiency.

CORGI The abbreviation for the Council for Registered Gas Installers. All gas fitting operatives must be registered with this organization in order to trade.

de-aerator See *air separator*.

direct cold water supply A cold water supply system that is fed directly from the mains supply pipe.

drain-off cock A small valve designed to permit a hosepipe to be connected and thereby allowing water to be drained from the system. All low points of water-filled systems should be fitted with a drain-off cock.

Essex flange A special fitting that is sometimes used to provide an additional connection to a hot water storage cylinder, such as where a shower requires its own supply.

f & e cistern The abbreviation for feed and expansion cistern. This is a small cistern used for supplying water to a central heating system. It also allows for the expansion of the water due to heating. When filled the water level is adjusted to a level very low down inside, just above the outlet point, thereby making room for the expanding water.

feed cistern A cistern located in the roof space to hold a quantity of cold water for the purpose of supplying a system of domestic hot water supply.

female iron A term used when referring to threaded joints. The female iron thread has an internal thread. The male iron has an external thread and screws into the female iron.

ferrous metal Metals that contain iron.

first fix A term used to indicate that the pipework has been run to an installation, but the appliances have yet to be fixed.

flame failure device A special control that prevents fuel passing to the combustion chamber of an appliance if a flame is not detected.

float-operated valve The control valve located inside a cistern to stop the incoming flow of water. It is often referred to as a ballvalve.

flow rate The volume of liquid that passes through the pipework.

flue The pipe that removes the products of combustion to the external environment.

flushing cistern Another name for a toilet cistern.

flux A special paste that is applied to the mating surfaces of copper pipe and fittings prior to soldering. It is designed to exclude the oxygen within the surrounding air which would otherwise cause the joint to become dirty when the heat is applied. It also helps the solder to stick to the pipe and fitting.

foul water The water from waste appliances and toilets.

fur A name sometimes used to refer to limescale.

gatevalve A stop valve which closes off the flow of water by closing a gate.

gland nut Also called a packing gland nut, it is the nut found on many taps and valves, designed to tighten up and squeeze out the material surrounding the spindle where it turns, thus preventing water leaking past this spindle. See figure 4.1.

gradient The incline of a pipe.

gully A drainage fitting into which smaller pipes are connected. The gully may be trapped or untrapped. All gullies connected to a foul-water drainage system must be trapped.

hard water A term used to describe water that has a proportion of calcium held in suspension within the water.

heat exchanger The component within a boiler or hot water cylinder where heat is transferred from one source to another.

hopper head A funnel-shaped drainage fitting that is located at the upper most end of a pipe. It is designed to assist the collection of water from smaller pipes which discharge into it.

immersion heater The heater found inside a hot water cylinder, similar to a kettle element, only much larger.

inhibitor A solution added to a central heating system, designed to minimize the problems of corrosion. It also has pump-lubricating qualities.

insulation The material applied to pipes and storage vessels to prevent the transference of heat or sound.

intercepting trap A special drainage fitting installed at the point where the house drain meets the sewer. It forms a trap to keep the two systems separate. These are no longer installed.

jointing paste One of several oil-based compounds that can be used to assist in making pipe connections. It is essential that this paste is not used in connections with plastic materials as it will cause them to break down.

jumper This is the brass plate onto which the washer of a tap is attached.

legionella A bacteria that grows in warm water and is potentially fatal when transmitted to humans via water in the form of a fine spray or mist.

LPG Abbreviation for Liquefied Petroleum Gas.

macerator pump A domestic pumping unit designed to discharge the low-level waste water contents from sanitary appliances through small-bore pipes up to a drainage system at a higher level.

male iron See *female iron.*

manhole An inspection chamber located below ground level, designed to give access to the house drains.

micro-bore Describes small-diameter central heating systems using pipes as small as 6 mm.

mixer tap A tap designed to receive both hot and cold water and deliver them into the appliance simultaneously, if desired.

motorized valve A special valve used within a fully pumped central heating system to automatically open and close the pipeline as and when the water is required for a particular circuit.

Multiquick The trade name of a push-fit WC pan outlet connector.

neutral point The point within a pumped heating system that is under the influence of atmospheric pressure and not subject to any positive or negative pressure caused by a pump.

'O' ring A neoprene washer, designed to prevent water escaping past two mating surfaces. These washers are used in push-fit joints and many taps and mixer valves to allow the turning movement of a spindle or spout.

open flue The flue pipe from a heat-producing appliance, in which the air supply to the appliance has been taken from the room.

overflow pipe The pipe found in toilet and storage cisterns, designed to remove the excess water filling the cistern when the float-operated valve fails to close off the water supply.

P trap A trap located on the outlet of a sanitary appliance, with its outlet horizontal to the floor.

Ph value Abbreviation for potential of hydrogen, identifying the amount of hardness in a sample of water.

pillar tap A design of tap that screws vertically into the sanitary appliance, such as might be used for a basin tap. See also *bib tap*.

pipe duct A void through which a pipe has been run, thereby facilitating maintenance or removal.

plumbers mait Non-setting putty often used when making the waste fitting connections to sanitary appliances.

PTFE tape Abbreviation for Polytetrafluoroethylene, which is a white coloured jointing tape that is generally used when making pipe threaded joints.

pump A device used to move a volume of water. One example would be a circulating pump used in a central heating system to force the water around the system.

PVC Abbreviation for polyvinyl chloride, which is a type of plastic, typically used in waste water drainage systems.

radiant heating A form of infrared heating, which is designed only to warm the objects upon which the heat waves land (typically the building structure or the occupants of the building). It does not warm the air.

radiator A heat emitter through which hot water is passed to warm a room.

reducer A pipe fitting designed to reduce or increase the bore of a pipe.

room-sealed appliance A fuel-burning appliance that takes its air supply from outside the building as well as discharging the flue products outside, usually at a point adjacent to the air intake, in which case it is referred to as a balanced flue. Where the flue discharges at a different location from the air intake it is not in balance.

running trap A trap which is installed within a run of pipework, rather than fixed directly onto a waste fitting, as found typically with the *'P'* or *'S' trap*.

sanitary appliance The name given to items such as baths, basins, toilets, sinks etc.

sealed system A central heating system that is not open to the atmosphere and typically is filled via a temporary filling loop connected to the water supply mains.

seating The area within a tap or valve onto which a washer is tightened, thus closing the supply.

separate system of drainage A system of drainage that has two drainage pipes, one for the surface rainwater and one for the foul water from the house sanitary pipework.

septic tank A private sewage disposal system used in some rural areas.

silicone A non-metallic substance, typically found in liquid form as a lubricant or as a rubber setting multipurpose sealing compound.

siphonic action The action of transferring a liquid, typically water, from one level up and over the edge of a vessel down to a lower level, using the force created by atmospheric pressure.

soft water A term used to describe water that has no calcium held in suspension within it, and which may contain additional carbon dioxide, making it more acidic.

solder An alloy used to make capillary joints in copper tube. The solder may contain lead and tin or it may be lead free, typically consisting of copper and tin.

stopcock A valve fitted on high pressure mains supply pipework, designed to control or stop the flow of water.

stop valve A valve installed within a run of pipe, e.g. a stopcock, gatevalve or quarter-turn.

storage cistern A large open-topped vessel, usually located in the roof space or loft, designed to hold water to supply a system of hot or cold water.

S trap A trap located on the outlet of a sanitary appliance with its outlet vertical to the floor.

stuffing box The area beneath the packing gland nut where a special packing is located to allow the spindle of a valve to turn without letting water escape. See *gland nut*.

Supa-tap A design of tap that can be re-washered without turning off the water supply.

surface water Rainwater that runs off from roofs and paved areas.

thermostat A device designed to automatically open or close the electrical circuit as the temperature increases or decreases.

trap A component located beneath or forming part of a sanitary appliance, designed to hold a quantity of water with the intention of preventing foul air and gasses passing into the building from the house drains.

unvented domestic hot water supply A system of stored hot water supply, in excess of 15 litres, that is supplied directly from the mains supply pipework.

vacuum A space devoid of any matter, therefore air from the surrounding atmosphere applies a force as it tries to fill the void.

valve A fitting incorporated within a pipe run, either to control the volume of liquid passing through or to completely stop the flow.

vent pipe The pipe from a high spot within a central heating or hot water system, designed to allow the air to escape as the system fills and let air into the system as it is drained.

washing machine trap A special trap with a branch for the connection of a dish washer or washing machine.

water pressure The force acting upon the water within the pipe. See page 13.

WC The abbreviation for water closet, which is a toilet. The term WC really identifies the room in which a toilet is located, but it is also used when referring to the toilet itself.

zone valve A two-port motorized valve used in a heating system to open and close the pipeline automatically when called upon by the thermostat.

Further reading

Treloar, R. D. (2006) *Plumbing* (third edition), London: Blackwell Publishing

Treloar, R. D. (2003) *Plumbing Encyclopaedia* (third edition), London: Blackwell Publishing

Treloar, R. D. (2005) *Gas Installation Technology*, London: Blackwell Publishing

Plumbing trade and professional bodies

Listed here is a selection of organizations from which the contact details of qualified operatives can be sought. The companies listed with organizations such as these will need to follow strict guidelines as laid down by the organization making the recommendation and, as such, the organization will be held accountable to some extent for the work that they undertake.

Association of Plumbing and Heating Contractors (APHC)
024 7647 0627
www.competentpersonsscheme.co.uk/consumers

Institute of Plumbing and Heating Engineering (IPHE)
01708 472791 www.iphe.org.uk

Council of Registered Gas Installers (CORGI)
0870 401 2300 www.trustcorgi.com

Oil Firing Technical Association (OFTEC)
0845 65 85 080 www.ofteconline.com

In addition, the following sponsored government website is very useful: **www.trustmark.org.uk**

index

Bridge
British Citizenship Test, The
British Empire, The
British Monarchy from Henry VIII, The
Buddhism
Bulgarian
Bulgarian Conversation
Business French
Business Plans
Business Spanish
Business Studies
C++
Calculus
Calligraphy
Cantonese
Caravanning
Car Buying and Maintenance
Card Games
Catalan
Chess
Chi Kung
Chinese Medicine
Christianity
Classical Music
Coaching
Cold War, The
Collecting
Computing for the Over 50s
Consulting
Copywriting
Correct English
Counselling
Creative Writing
Cricket
Croatian
Crystal Healing
CVs
Czech
Danish
Decluttering
Desktop Publishing
Detox
Digital Home Movie Making
Digital Photography
Dog Training
Drawing
Dream Interpretation
Dutch
Dutch Conversation
Dutch Dictionary
Dutch Grammar
Eastern Philosophy
Electronics
English as a Foreign Language
English Grammar
English Grammar as a Foreign Language
Entrepreneurship
Estonian
Ethics
Excel 2003
Feng Shui
Film Making
Film Studies
Finance for Non-Financial Managers
Finnish
First World War, The
Fitness
Flash 8
Flash MX
Flexible Working
Flirting
Flower Arranging
Franchising
French
French Conversation
French Dictionary
French for Homebuyers
French Grammar
French Phrasebook
French Starter Kit
French Verbs
French Vocabulary
Freud
Gaelic
Gaelic Conversation
Gaelic Dictionary
Gardening
Genetics
Geology
German
German Conversation

German Grammar
German Phrasebook
German Starter Kit
German Vocabulary
Globalization
Go
Golf
Good Study Skills
Great Sex
Green Parenting
Greek
Greek Conversation
Greek Phrasebook
Growing Your Business
Guitar
Gulf Arabic
Hand Reflexology
Hausa
Herbal Medicine
Hieroglyphics
Hindi
Hindi Conversation
Hinduism
History of Ireland, The
Home PC Maintenance and Networking
How to DJ
How to Run a Marathon
How to Win at Casino Games
How to Win at Horse Racing
How to Win at Online Gambling
How to Win at Poker
How to Write a Blockbuster
Human Anatomy & Physiology
Hungarian
Icelandic
Improve Your French
Improve Your German
Improve Your Italian
Improve Your Spanish
Improving Your Employability
Indian Head Massage
Indonesian
Instant French
Instant German
Instant Greek
Instant Italian
Instant Japanese
Instant Portuguese
Instant Russian
Instant Spanish
Internet, The
Irish
Irish Conversation
Irish Grammar
Islam
Israeli-Palestinian Conflict, The
Italian
Italian Conversation
Italian for Homebuyers
Italian Grammar
Italian Phrasebook
Italian Starter Kit
Italian Verbs
Italian Vocabulary
Japanese
Japanese Conversation
Java
JavaScript
Jazz
Jewellery Making
Judaism
Jung
Kama Sutra, The
Keeping Aquarium Fish
Keeping Pigs
Keeping Poultry
Keeping a Rabbit
Knitting
Korean
Latin
Latin American Spanish
Latin Dictionary
Latin Grammar
Letter Writing Skills
Life at 50: For Men
Life at 50: For Women
Life Coaching
Linguistics
LINUX
Lithuanian
Magic
Mahjong
Malay

Managing Stress
Managing Your Own Career
Mandarin Chinese
Mandarin Chinese Conversation
Marketing
Marx
Massage
Mathematics
Meditation
Middle East Since 1945, The
Modern China
Modern Hebrew
Modern Persian
Mosaics
Music Theory
Mussolini's Italy
Nazi Germany
Negotiating
Nepali
New Testament Greek
NLP
Norwegian
Norwegian Conversation
Old English
One-Day French
One-Day French – the DVD
One-Day German
One-Day Greek
One-Day Italian
One-Day Polish
One-Day Portuguese
One-Day Spanish
One-Day Spanish – the DVD
One-Day Turkish
Origami
Owning a Cat
Owning a Horse
Panjabi
PC Networking for Small Businesses
Personal Safety and Self Defence
Philosophy
Philosophy of Mind
Philosophy of Religion
Phone French
Phone German
Phone Italian
Phone Japanese
Phone Mandarin Chinese
Phone Spanish
Photography
Photoshop
PHP with MySQL
Physics
Piano
Pilates
Planning Your Wedding
Polish
Polish Conversation
Politics
Portuguese
Portuguese Conversation
Portuguese for Homebuyers
Portuguese Grammar
Portuguese Phrasebook
Postmodernism
Pottery
PowerPoint 2003
PR
Project Management
Psychology
Quick Fix French Grammar
Quick Fix German Grammar
Quick Fix Italian Grammar
Quick Fix Spanish Grammar
Quick Fix: Access 2002
Quick Fix: Excel 2000
Quick Fix: Excel 2002
Quick Fix: HTML
Quick Fix: Windows XP
Quick Fix: Word
Quilting
Recruitment
Reflexology
Reiki
Relaxation
Retaining Staff
Romanian
Running Your Own Business
Russian
Russian Conversation
Russian Grammar
Sage Line 50

Sanskrit
Screenwriting
Second World War, The
Serbian
Setting Up a Small Business
Shorthand Pitman 2000
Sikhism
Singing
Slovene
Small Business Accounting
Small Business Health Check
Songwriting
Spanish
Spanish Conversation
Spanish Dictionary
Spanish for Homebuyers
Spanish Grammar
Spanish Phrasebook
Spanish Starter Kit
Spanish Verbs
Spanish Vocabulary
Speaking On Special Occasions
Speed Reading
Stalin's Russia
Stand Up Comedy
Statistics
Stop Smoking
Sudoku
Swahili
Swahili Dictionary
Swedish
Swedish Conversation
Tagalog
Tai Chi
Tantric Sex
Tap Dancing
Teaching English as a Foreign Language
Teams & Team Working
Thai
Thai Conversation
Theatre
Time Management
Tracing Your Family History
Training
Travel Writing
Trigonometry
Turkish
Turkish Conversation
Twentieth Century USA
Typing
Ukrainian
Understanding Tax for Small Businesses
Understanding Terrorism
Urdu
Vietnamese
Visual Basic
Volcanoes, Earthquakes and Tsunamis
Watercolour Painting
Weight Control through Diet & Exercise
Welsh
Welsh Conversation
Welsh Dictionary
Welsh Grammar
Wills & Probate
Windows XP
Wine Tasting
Winning at Job Interviews
Word 2003
World Faiths
Writing Crime Fiction
Writing for Children
Writing for Magazines
Writing a Novel
Writing a Play
Writing Poetry
Xhosa
Yiddish
Yoga
Your Wedding
Zen
Zulu

teach yourself

basic DIY

DIY doctor

- Do you want a clear, concise guide to DIY essentials?
- Would you like advice on repairs and maintenance?
- Do you need help undertaking new projects in your home?

Basic DIY explains the essentials of decorating and maintenance in a step-by-step manner. With practical advice, clear illustrations and plenty of useful tips, it covers all the key skills, from carpentry to plastering, giving you the confidence to tackle the jobs you always thought you'd have to pay for.

DIY Doctor is a unique and hugely popular website, offering free self-build and DIY help from qualified tradespeople through an interactive question and answer service.

teach yourself

ethical living

peter macbride

- Would you like to have a more ethical lifestyle?
- Do you want to make more informed shopping decisions?
- Would you like realistic advice for a greener home or holiday?

Ethical Living is a realistic introduction to a more environmentally aware lifestyle. It gives you straightforward advice on making informed decisions about what to buy, how to travel and what you can and can't recycle. It has practical tools like calculators and website guides, and plenty of further resources to help you make positive and lasting changes in all the important areas.

Peter MacBride is a professional writer and researcher, and the author of over 120 books.

teach yourself

saving energy in the home

nick white

- Do you want to cut your energy bills?
- Would you like a more energy-efficient home?
- Do you want a greener life at little extra cost or effort?

Saving Energy in the Home offers you straightforward and achievable strategies for reducing your energy bills and living a more environmentally aware life. With lots of useful tools to assess your energy and carbon use, it gives practical advice on everything from heating to HIPs and sustainable gardening and even offers guidelines on generating your own power.

Nick White is a director and member of the Hockerton Housing Project – the UK's first earth sheltered, self-sufficient ecological housing development and the winner of several energy efficiency awards.

teach yourself

basic gardening skills

jane mcmorland hunter & chris kelly

- Are you clueless about where to start in your garden?
- Do you need to know what to do and when?
- Would you like guidance on growing your own vegetables?

Designed for the complete beginner, **Basic Gardening Skills** shows you how to turn a patch of muddy ground into an easily maintainable garden, whatever the size of your plot and however busy you are. From lawn care and watering to creating patios and growing vegetables, it is packed with easy-to-follow, practical advice.

Jane McMorland Hunter is a professionally trained designer and published author. **Chris Kelly** is a professional gardener and writer. Both have many years' experience designing, constructing and maintaining a variety of gardens.